UNDERSTANDING CREATION AND EVOLUTION: A BIBLICAL AND SCIENTIFIC COMPARATIVE STUDY.

Understand how God's creation of Man, as revealed in Genesis, and how today's science and evolutionary knowledge do not have to disagree.

- **If intelligently understood.**

Understand how descendants of Adam and Eve evolved into the races of Man, and what makes people different, and why it is so.

- **If Intelligently understood.**

By Howard Ray White

I0171208

Copyright: 2003, 2010 and 2014.

Introduction.

Understanding how th concept of Creation and the concept of Evolution relate — one to the other — seems to become more and more perplexing throughout the world as each decade comes and goes. I was born in 1938. My goodness, just consider the advances in science over the intervening 75 years I have witnessed, especially in human knowledge of the physical sciences and biology. What about the advances in human knowledge over the 200 years since my great, great grandfather was born?

But what about human Biblical knowledge?

For over 2,000 years God's revelations, through His prophets, have stood like a rock — not changing. On the other hand, over that span of over 20 centuries, human scientific knowledge has grown exponentially by leaps and bounds. So, can we students of Creation and Evolution make sense of the apparent ever-widening disagreement between these two foundations of human knowledge — foundations upon which human civilization has been and is continuing to be built.

So that was my dilemma — a dilemma that pushed me to engage in personal study aimed at seeking an understanding.

After considerable personal research and study, I take comfort in finding a pathway that makes sense of both Creation and Evolution, whereby their apparent conflicts are resolved and seem to fade away.

The pathway to the understanding we humans seek is revealed, to a considerable extent, by our realizing that God's revelation of His Creation is not, in fact, unchanging. It too is moving with time. Why? Because, God, it seems obviously apparent, had never revealed scientific knowledge to his prophets. Yes, God made Man with the intelligence and ability to, collectively, discover scientific truths and exploit them to advance human civilization and quality of life. And — this is so important — He has always been determined to present revelations to His prophets in language and concept that fits into each prophet's level of scientific, geographic and biological knowledge. What does that mean?

It means that words and concepts given to His prophets and passed down to us in today's Bible were chosen to fit within the understood terminology of the era when first revealed. And, it is that long-ago, primitive terminology that clouds our ability to understand His revelations. So, our challenge is to mentally place ourselves in the knowledge basis of thousands of years ago and seek to make sense of the book of Genesis from that perspective.

This is what I have striven to do. I have placed myself, mentally, into the knowledge basis of thousands of years ago, and then begun to read. In so doing, I have found a pathway that leads to understanding how both Creation and Evolution can be understood with far more agreement than generally thought.

And what about Evolution? How about human Evolution?

I have studied that as well. And I find that Evolution will become far more understandable to us today if we mate our Creation knowledge to our Evolution knowledge and examine the resulting child, which I will call "unified knowledge." My study attempts to do this — to resolve a "unified knowledge" that is derived from an **intelligent and unbiased analysis** of what we humans know about both Creation and Evolution.

This concludes my introduction. We now arrive at the presentation of my study. But before becoming immersed in the details, let me present a broad outline of the understanding attainable through subsequent pages:

1. You will understand that God's creation of Man, as revealed in Genesis, and today's science and evolutionary theory, do not disagree if intelligently understood.

   - You will understand how Genesis and science do not disagree, beginning with the creation of the Universe, and continuing through the creation of man (Adam and Eve).

   - You will understand how God employed both natural and supernatural powers to create man in such a way that he was, intentionally, compatible biologically with earthly life, but also possessed a soul capable of everlasting life.

2. You will learn how, following the era of Adam and Eve, man proliferated and struggled to survive and to cope with ever-present dangers, not the least of which was the horrific Ice Age, which drove into caves those who had bravely ventured north during earlier warmer years.

- You will learn how the contrast between the pleasant weather in Africa and the harsh, northern Ice Age weather caused man to evolve into two major classifications of race – the African races and the Out-of-Africa races, which we know broadly as European and Asian.

3. You will understand the four major differentiating characteristics of the major races of man that we recognized today, why those differentiations exist, and how those differentiations are now expressed.

In other words, you herein will learn how descendants of Adam and Eve evolved into the races of man, what makes individuals different, what makes races different, and why it is so.

Although the findings in this study are not politically correct, I hope you find them helpful, going forward, as all the good people of our nation seek to improve the lives of all Americans.

One final bit of advice: my study is irrevocably dedicated to seeking the truth, in accordance with the directive by Jesus Christ, who instructed:

In all endeavors seek the truth – *"for the truth shall set you free."*

Best wishes to you, and thanks for reading this study.

Howard Ray White.

Table of Contents

Let Us Begin Our Creation and Evolution Study.

I will present information about Creation and Evolution, based on the Bible and scientific studies, to encourage you to think in terms of the big picture. I will first cover the creation of the Universe and Earth. I will cover the creation of life. Then, I will cover God's creation of man in east Africa, followed by man's eventual migration out of Africa and into the various continents of earth. I will explain how, trapped and isolated by Ice Age weather, man thusly evolved into notable individual diversity and into the four major races of man. I will follow God's relationship with man through His seven covenants, ending in the covenant under which we live today. I will explore the four major differentiating characteristics of man, which differ slightly among individuals (even among siblings) and slightly, but noticeably, among the races.

Everything presented in my study is intended to encourage you to think about fundamental moral and religious issues related to every man's relationship to God, and every man's relationship to his fellow human beings, here among Earth's diverse population.

First, I should tell you that I am a Christian man. I believe that God created the Universe and the Earth and the life that lives upon it. I believe that God carried out that creation by employing the cosmic forces and evolutionary life forces that are observed by scientists today. But, I believe, God intervened supernaturally to create man in his own image – man did not evolve from animals merely through natural evolution.

And, from that basis, I will proceed.

All people today are descended from God's creation of Adam and Eve: a supernatural event. We are not just highly evolved animals. Unlike animals, each person has the gift of a soul that is capable of living an eternal life in the presence of God. And, through his son Jesus Christ, God has revealed today's path to Heaven: "*For God so loved the world that He gave His one and only Son, that whoever believes in Him shall not perish, but have everlasting life.*" Through Jesus Christ we are told: "*by grace you have been saved, through faith.*" And, that concept is so often quite contrary to man's common sense notion about how God ought to select his harvest of souls. So, to counter our common sense notions, God clearly warns that salvation is, "*not by works, so no one can boast.*"

Let us deal first with the issue of Creation versus Evolution up to the arrival of Man. Did the Universe just happen, or is it God's creation? Did man just evolve from animals, or did God rather recently create man in his image? I begin with my understanding regarding God's creation of the Universe and the Earth with its diverse plants and animals.

Chapter 1. God's Creation of the Heavens and Earth.

In the beginning, I have come to understand, God created the Universe. Astrophysicists tell us that the enormous Universe they observe from earth today is probably the result of a "Big Bang" in which mass and energy was hurled from an unbelievably dense mass that had been located at the center of the universe. And, based on the speed at which heavenly bodies appear to be flying apart today, astronomers calculate that the "Big Bang" event had taken place 12,000,000,000 years ago. God confirms, through his prophet, that He was the Creator of what we call the "Big Bang." The first part of the first sentence in the Torah, as reproduced in the Holy Bible, reports:

> *In the beginning God created the heavens . . .*

As the swirling matter and energy sped outward from the "Big Bang," and cooled, it subdivided into rotating regions that further gathered into rotating galactic discs, which we call galaxies. The swirling matter and energy that was hurled in our direction by the "Big Bang" gathered into the Milky Way Galaxy, and that is the galaxy in which we live. Within the Milky Way Galaxy, matter further cooled and gravitational forces gathered that matter into thousands of rotating stellar disks. In many of those rotating stellar disks, sufficient matter eventually gathered at the center to ignite thermonuclear fusion, and that caused the central mass within that disk to shine as a star.

Perhaps I should digress for a moment and tell you how many stars we now believe presently exist in the Universe. Then I will present an estimate of how many grains of sand presently exist in all of the beaches on Earth. Then I will present the United States national debt in dollars. You will find these numbers of interest, for it is the size of God's humongous Universe that it is so hard for little human beings like us to comprehend.

- Total number of stars in the Universe: 300,000,000,000,000,000,000,000

- Grains of sand on all of the beaches on Earth: 4,800,000,000,000,000,000,000

- The U. S. National Debt in dollars (October 12, 2013): 16,750,794,000,000

Returning to a single star, such as our Sun, we find that, stair-stepping outward from the central mass within each rotating stellar disk, matter was also gathering into an array of primitive planets. God confirms through his prophet how Earth looked at this point. The complete first and the second sentence in the Bible reports:

> *In the beginning God created the heavens and the earth. Now, the earth was formless and empty, darkness was over the surface of the deep, and the Spirit of God was hovering over the waters.*

Since language was limited by primitive scientific understanding at the time God delivered this revelation, *"waters"* was perhaps the word chosen to represent all "fluids," for the heat within the forming mass, which was to become Earth, certainly made that mass a very hot fluid environment. Our central mass was destined to become our "sun", but it had not yet triggered thermonuclear fusion, and clouds of dust and gases on primitive "Earth" obscured the light from distant stars. So *"darkness was over the surface of the deep."*

So the forming primitive planets revolved around our central mass. Also, most planets spun as they revolved around our central mass. Planets had insufficient mass to be stars, but thermonuclear fission, gravitational forces and meteor and comet impacts, caused the newer planets to be very hot and volcanic. It would take a long time for meteor and comet impacts to subside and for planets to cool down enough to retain much water or a friendly atmosphere. Unable to remain at the surface of turbulent planets, incoming water was pushed back out into space, where it spent long periods in the form of ice, and often gathered into comets, or it became chemically bound within earth's rock. Now, at some point our central mass became dense enough to support thermonuclear fusion, and a star was born. And each spinning planet exposed to that new sunlight experienced sunrises and sunsets. Our central mass, the star that gives us life, became what we call, the "Sun". Astrophysicists calculate that the Sun began to shine 6,000,000,000 years ago. The planet on which we live, we call, the "Earth". God confirms through his prophet how Earth looked at this point. The third through the sixth sentences in the Bible report:

> *And God said, "Let there be light," and there was light. God saw that the light was good, and he separated the light from the darkness. God called the light "day," and the darkness he called "night". And there was evening, and there was morning – the first day.*

We now move into "the second day," still mentally limited in our Biblical mind-set to a primitive knowledge. As Earth cooled further, it acquired an atmosphere of nitrogen and carbon dioxide. Initially, there was very little water on Earth, for most existed out in the space occupied by the Sun and her planets, of which Earth was number three. So, there was an expanse of atmosphere, and, beyond that, an expanse of vast space, where ice roamed. Earth was still harsh, and without life. But, the ice attracted to Earth was no longer being pushed back into space, for the Earth had cooled sufficiently for liquid water to accumulate on its surface, and for ice to accumulate at its poles. So, there was water on Earth, and an atmosphere above. But, most of the water we see today on Earth was still out in space, in the form of roaming ice. God confirms through his prophet how Earth looked at this point. The seventh through the eleventh sentences in the Bible report:

> *And God said, "Let there be an expanse between the waters to separate water from water." So, God made the expanse and separated the water under the expanse from the water above it. And it was so. God called the expanse "sky". And there was evening, and there was morning – the second day.*

Again, the language available to God's prophet limited how the story of the atmosphere could be revealed. There was no word to distinguish "space" from "atmosphere". So, everything above the ground was "sky".

The presence of water on Earth accelerated the cooling of the surface. As it further cooled, the Earth's crust fractured into tectonic plates, and those plates began to slowly move across the molten core within. New crust steadily emerged through trenches in the young oceans, forming new crust at that edge of the tectonic plate. And, old crust steadily melted away where it dove down at the far edge, where two adjacent plates moved toward each other with terrific force. So, fresh land was continually being formed beneath the young oceans, and was being conveyed by interior forces toward the beaches, and beyond, where it became new dry land. And, the oceans became deeper and more defined, as crust movement became more defined, and as more ice was swept into earth's gravitational field from the space through which the Earth roamed. The action of water and wind erosion and the emergence of land from beneath the oceans produced a primitive soil. God confirms through his prophet how Earth looked at this point. The twelfth through the fifteenth sentences in the Bible report:

> And God said, "Let the water under the sky be gathered to one place, and let dry ground appear." And it was so. God called the dry ground "land", and the gathered waters he called "seas". And God saw that it was good.

Chapter 2. God's Creation of Life.

Applying knowledge based on today's astronomical knowledge, I must rationalize that throughout the Universe there are nurturing planets and moons where life abounds, millions of them (The Bible does not deny the existence of life beyond elsewhere). One of these is Earth. It was about 4,000,000,000 years ago, when primitive life began to arrive at Earth and survive, multiplying and replicating itself according to its kind. These primitive life chemicals were arriving in the ice that Earth continually swept from space, especially within an occasional ice-covered meteor, or, much less frequently – within an arriving comet. The first life was much more primitive than today's viruses or bacteria; it was the chemical building blocks of life – what we call RNA. RNA survives today as a component within each cell in our body. These first life forms lived under the ice fields of the oceans and drew energy from geothermal flumes on the ocean floor below. The temperature and crystal structure of the impure ice sustained the fragile chemicals of primitive life forms, while the geothermal flumes provided nutrients and energy. Overhead, the atmosphere was almost totally made up of carbon dioxide and nitrogen. There was no significant oxygen. And the clouds and atmospheric dust made every day look like a dim cloudy winter day. If we had visited earth at that time, we would not have been able to see anything overhead but dense clouds: no Sun, no Moon, and no stars. As time passed, life within the ocean became more complex and viruses and bacteria appeared. Yet no life form drew energy from oxygen, for little was present. In fact, life struggled to develop defenses against being destroyed by the little oxygen that did exist, for its presence was increasing. Photosynthesis had begun. Using the dim sunlight that was penetrating down to the surface of the oceans, some types of single-cell organisms were converting carbon dioxide to oxygen. Yet, there was nothing as complex as the smallest minnow, snail or sea-worm. The most successful life forms were the large humps of stromatolite, where colonies of diverse microbes thrived by symbiotic relationships.

But, life did not remain in the ocean, because the ocean floor was continually being conveyed up and across the beaches by tectonic forces within earth's interior. By riding upon the emerging ocean floor, life came to the dry land. Exposure to the dryness of land killed almost all of the life soon after it emerged from the ocean, but a remnant survived, and strengthened its ability to cope in that harsh environment. In time, these ocean plants developed into primitive land plants. Now, forces were in place to accelerate changes in the earth's atmosphere. No longer would the atmosphere consist of carbon dioxide and nitrogen, with little oxygen, for much oxygen was being produced by the new land plants and the expanded life forms within the ocean waters. The sixteenth through the nineteenth sentences in the Bible report:

And God said, "Let the land produce vegetation: seed-bearing plants and trees on the land that bear fruit with seed in it, according to their various kinds." And it was so. The land produced vegetation: plants bearing seed according to their kinds, and trees bearing fruit with seed in it according to their kinds. And God saw that it was good. And there was evening, and there was morning – the third day.

By 2,000,000,000 years ago, there was significant oxygen present in the atmosphere. This made possible a new form of life: life that derived its energy by consuming oxygen. Oxygen-consuming life forms would complement the carbon dioxide-consuming life forms that were already abundant, and thereby complete a vital cycle in which essential life-supporting chemicals would never be depleted. As the atmosphere enriched in oxygen, and the earth matured, the perpetual cloudiness changed to occasional cloudiness. Had man been present on earth at this time, he could have looked up and, for the first time, seen the sun, the moon and the stars. And, God revealed to His prophets the story of this step in His sequence of creation, but language problems must have confounded the story. (Remember our guiding premise: we must recalibrate our minds back to the primitive knowledge platform that was prevalent thousands of years ago.) Instead of the account being passed down through the generations as "God made the sun, moon and stars visible in the sky," the account was passed down as *God made the sun, moon and stars and placed them in the sky.*" The twentieth through the twenty-sixth sentences in the Bible report:

> *And God said, "Let there be lights in the expanse of the sky to separate the day from the night, and let them serve as signs to mark seasons and days and years, and let there be lights in the expanse of the sky to give light on the earth." And it was so. God made two great lights – the greater light to govern the day and the lesser light to govern the night. He also made the stars. God set them in the expanse of the sky to give light on earth, to govern the day and the night, and to separate light from darkness. And God saw that it was good. And there was evening, and there was morning – the fourth day.*

At this point Earth had a viable regenerative life cycle, which was based on the vital synergism between plants and animals. The inexhaustible carbon dioxide-to oxygen-to carbon dioxide cycle was now thriving. So, new life expanded by leaps and bounds. Occasionally, new ice-bound DNA arrived in meteors and comets, survived the plunge through earth's atmosphere, and further seeded new and unique life forms. Mutations expanded the variations within established life-form families. New life forms proliferated, most noticeably in the oceans, where life had originated. In the ocean, there was a rich base of living single-cell life forms on which to base an upward spiraling food chain. So, various fishes and seabirds evolved according to God's plan. Life in the ocean was no longer tiny creatures too small to be seen without a microscope. By 500,000,000 years ago the ocean had become home to plants and animals in sizes similar to the more diminutive species we see today. At the height of this era, while nesting on land out of reach of ocean-bound predators, flying reptiles fed off the life forms within the oceans with abound. The twenty-seventh through the thirty-first sentences in the Bible report:

> *And God said, "Let the water teem with living creatures, and let birds fly above the earth across the expanse of the sky." So God created the great creatures of the sea and every living and moving thing with which the water teems, according to their kinds, and every winged bird according to its kind. And God saw that it was good. God blessed them and said, "Be fruitful and*

increase in number and fill the water in the seas, and let the birds increase on the earth." And there was evening, and there was morning – the fifth day.

However, some birds remained on the land and developed the ability to feed off the growing supply of land plants. And, they went inland to lakes and established themselves there. Other sea life evolved the ability to crawl, then to walk upon the land, and to migrate far inland. By this evolution, God populated the land with many kinds of animals, from small to large. The dinosaurs arose about 200,000,000 years ago, but suddenly died off 65,000,000 years ago, victims of an asteroid or comet collision with Earth. Yet, that global catastrophe encouraged the evolution of the mammal species we see today. Some animals fed on plants. Among those would be the animals that man would eventually domesticate. Others fed on other animals. Some fed on both. The thirty-eighth through the forty-first sentences in the Bible report:

And God said, "Let the land produce living creatures according to their kinds: livestock, creatures that move along the ground, and wild animals, each according to its kind." And it was so. God made the wild animals according to their kinds, the livestock according to their kinds, and all the creatures that moved along the ground according to their kinds. And God saw that it was good.

Chapter 3. God's Creation of Man.

So, God had created the heavens and the Earth, and He had created the plants and animals living upon the Earth. This He revealed through His prophets and the account has been passed down to us through the first forty-one sentences in the Bible. God presented His account in terms of six "phases," which His prophets symbolized as six "days." The sixth phase yet remained. It would be the creation of man.

Because it is illogical to think otherwise when considering the vastness of the Universe, my understanding is that, throughout the Universe, there are planets and moons where advanced plant and animal ecosystems flourish. I further suspect and submit to you that, when one of these advanced nurturing planets or moons became capable of supporting a being made in the image of God, He quite likely proceeded to create a seed community of intelligent beings, each complete with a supernatural soul, and each customized to live in the environment of that site. But that is conjecture. We must confine our search for understanding closer to home.

It was about 200,000 years ago that God, upon recognizing that Earth was at this point in evolution, chose to create man.

It is apparent today from fossil records that, for about 2,000,000 years, primate animals and animals that superficially resembled man had been living on earth. Some of the animals that resembled man, which I will call hominid animals, even used crude stone tools. Apparently, the most advanced hominid animals and primate animals lived in east Africa. But these were all animals, I believe; the creation of man would be God's handiwork. Let me postulate how it might have been done:

Wisely desiring that man's body share the biology of earthly animals, I suppose that God created man by modifying the DNA structure of a hominid animal or sub-human animal instead of starting anew. As evidence supporting this supposition, I remind you that the nuclear DNA of Neanderthals living 40,000 years ago matches 99.7 percent of the nuclear DNA of modern man, and that today's chimpanzee and man today share 98.8 percent of the same nuclear DNA structure. I have no insight into how God created the natural side of man, but suggest that He might have inserted man's DNA structure into a barren egg retrieved from the chosen primate or hominid surrogate mother (today's biotechnology science points to that technique). It seems logical for God to use the biochemistry that had evolved on earth, for man was to be compatible with earth's eco-system with regard to diet and resistance to environmental and biological hazards. Furthermore, I suppose God had no reason to want man to be unnecessarily different from the animals on earth with regard to biology. That would have served no useful purpose it seems to me. So, I suggest that God took an existing hominid or primate animal DNA structure and modified it to create an intelligent, athletic being capable of living anywhere on earth, whether hot or cold, dry or wet – a being remarkably advanced in intelligence from any animal that had ever lived on earth – a being capable of elaborate speech and beautiful music – an energetic being with immense capacity for love and hate – but a being whose fragile childhood body was extremely dependent on many years of attentive nurturing. In addition, I believe God gave man something unique and even more remarkably special beyond the biological gifts. I believe God gave man a spirit capacity called a soul, for He wanted to be able to later invite the souls of chosen men, women and children into Heaven.

Again, I remind you that I am presenting my belief to stimulate your thinking on these matters. That God created man and man's soul I am personally convinced. Although I have only guessed at how he accomplished the natural side of that feat, I have no idea how he gave man a supernatural soul.

So, by whatever method used, in due time, a baby boy was born: the first human. God gave him the name "Adam." I would expect that Adam more closely resembled the babies born to African women today than any other modern race. God kept close watch over little Adam. Before many years had passed, God took a few cells from Adam and created the egg for the first female "man," and, if my previous reasoning is applied, He would have planted it into a chosen female hominid or primate's womb. That egg grew into the female "man," which God named "Eve." As with Adam, God would look over the rearing of Eve. And, with Adam as the father, Eve would later bear children. The creation of baby Adam and baby Eve, and their rearing and training through both childhood and adulthood, represented God's First Covenant with man, which I will term, God's "*Covenant of Nurture.*"

Chapter 4. God's First Covenant: His Covenant of Nurture — Adam and Eve.

God would later reveal to man, through his first prophets, probably in the days of Noah, that He had named the site of man's creation "Eden," and that He had named the first male, "Adam," and the first female, "Eve." Since the prophets had no knowledge of Africa, which was the true first homeland of early man, the account, particularly during its retelling down through subsequent generations, became identified as being on the east bank of Black Lake – later to become today's Black Sea – which was the home of Noah and his people. Students of geography will recall that the Black Sea, a large body of salt water, is north of Turkey and south of Ukraine. Since the prophets had no knowledge of cells and DNA, they would describe that microscopic material as *"dust of the ground,"* and they would describe the removed structural piece of Adam's body as a *"rib"*.

What the prophets did not say is important. Yes, it is important that we understand that the prophet's presentation of God's creation of man did not make any logical sense to ancient believers and only makes sense to modern-day scientists. The prophets did not say that God made a likeness of Adam or Eve out of molded clay or carved stone, and then breathed life into it; did not say that He transformed some animal into Adam and Eve, et cetera. Amazingly, the prophets described DNA as the smallest known particle of matter, a particle so small that it seemed to float in the air – dust. And they **did** choose to tell that Eve was created from a small, dispensable part of Adam's body – a rib. Was this a biological sample of genetic bone or soft tissue? Is the story about surgery? It seems to be a story out of a modern genetics fertility laboratory.

God's prophet reveals the story of God's creation of Adam and Eve in the following way:

> Then, God said, "Let us make man in our image, in our likeness, and let them rule over the fish of the sea and the birds of the air, over the livestock, over all the earth, and over all the creatures that move along the ground." So God created man in his own image, in the image of God he created him; male and female he created them.

> So, the Lord God caused the man to fall into a deep sleep; and while he was sleeping, he took one of the man's ribs and closed up the place with flesh. Then the Lord God made a woman from the rib he had taken out of the man, and he brought her to the man.

> Adam named his wife Eve, because she would become the mother of all the living.

> The man said,

> "This is now bone of my bones
> and flesh of my flesh;
> she shall be called 'woman,'

for she was taken out of man."

God blessed them and said to them, "Be fruitful and increase in number; fill the earth and subdue it. Rule over the fish of the sea and the birds of the air and over every living creature that moves on the ground."

Then, God said, "I give you every seed-bearing plant on the face of the whole earth and every tree that has fruit with seed in it. They will be yours for food. And to all the beasts of the earth and all the birds of the air and all the creatures that move on the ground – everything that has the breath of life in it – I give every green plant for food." And it was so.

God saw all that he had made, and it was very good. And there was evening, and there was morning – the sixth day.

During the *Covenant of Nurture*, Adam and Eve's clan grew to several generations. Yet, it seems obvious to me, they continued to be totally dependent on God and his angels for their survival, because they had no one else to teach them coping skills. One can imagine that members of the clan were bonded to God and his angels in a manner resembling the way a servant might have been bonded to his master a few hundred years ago. The servant had limited freedom and, in exchange, the master was totally responsible for the welfare of the servant and his family. During this formative period, men and women were obligated to honor God, to be subservient to God, and follow the rules of conduct God had taught them. They were expected to suppress creative thinking about what ought to be considered "good," and what ought to be considered "evil." God taught Adam's clan the beginnings of language. Perhaps, God began that process by first encouraging man to give names to the plants and animals he encountered, and to remember those names. Perhaps God taught man to make fire and cook meat. He probably taught man how to build crude huts, and how to make clothing from animal skins and from woven grasses, tied with vines or strips of bark. Surely, God taught man how to make spears and club weapons to enable males to defend the clan from wild animals and hominid animals. Yes, it seems apparent that God had a lot invested in Adam and Eve and their clan, and He realized that, at the beginning, man was perhaps the most fragile being on earth. If Adam's clan was killed, God knew that He would have to start over by creating another baby Adam.

God's prophet reveals to us the story of how God used the creation site in east Africa, Eden, to nurture Adam, Eve and their immediate descendants. (Modern scientific exploration makes us comfortable in placing Eden in East Africa.) But, where did His prophet place Eden on the map of the known world. It was not placed in Africa, because that region was not known to the people to whom the prophet was speaking. No, Eden was confused with geography closer to home.

A few pages further in this study will bring you to the story of Noah and the Great Flood. We have a preview of that story just below, where you read about four noteworthy rivers that are associated with Eden. After reading the account of Noah a few pages later is this study, you might want to return here to the story of Eden for a second interpretation. If so, then consider the possibility that God's prophet and later generations confused the actual site of Adam and Eve's homeland with Noah's original homeland on the shores of Black Lake, now the Black Sea, and with Abraham's homeland in the Tigris and Euphrates Valley. Perhaps, the Gihon, a word meaning "spurter," was a river feeding Black Lake, such as the Danube, and perhaps the Pishon, a word meaning "gusher," was the Bosporus floodwater pouring into Black Lake. Perhaps, the Tigris and Euphrates rivers were added to the story after the catastrophic flooding had driven Black Lake people to Mesopotamia. Perhaps, the four rivers symbolize a family tree, a history of a people, whereby the ancestral rivers, Gihon and Pishon, flow forward in time, to create the descendant rivers, Tigris and Euphrates. I am sorry if this is confusing. It will make more sense later. The prophets' account of Eden begins.

> *Now the Lord God had planted a garden in the east of Eden; and there he put the man he had formed. And the Lord God made all kinds of trees grow out of the ground – trees that were pleasing to the eye and good for food. In the middle of the garden were the tree of life and the tree of the knowledge of good and evil.*

> *A river watering the garden flowed from Eden; from there it was separated into four headwaters. The name of the first is the Pishon; it winds through the entire land of Havilah, where there is gold. (The gold of that land is good; aromatic resin and onyx are also there.) The name of the second river is the Gihon; it winds through the entire land of Cush. The name of the third river is the Tigris; it runs along the east side of Asshur. And the fourth river is the Euphrates.*

> *The Lord God took the man and put him in the Garden of Eden to work it and take care of it. And, the Lord God commanded the man, "You are free to eat from any tree in the garden; but you must not eat from the tree of the knowledge of good and evil, for when you eat of it you will surely die."*

After a few generations were born, and Adam's clan had attained a size and a level of skill that God believed would be sufficient for man-kind's survival, He prepared to set man free from such close oversight. Now, being free meant that God and His angels would not be hovering nearby ready to rush in to deflect danger from outside the clan, to settle an internal squabble within the clan, to circumvent a food shortage, to deal with a severe cold spell, or to help rebuild a hut. Man would have to grapple with good and evil by coping with issues himself. Supernatural help from God and His angels would be much more remote.

It seems to me that God wanted to stress this change in the relationship between Himself and man in a symbolic way. The story of that symbolism is handed down, by God's revelation through his prophets, as the story of the tree of good and evil that grew in the center of a garden that God had prepared in the eastern part of a region named "Eden." A quick read of that story implies that the first man, Adam, and the first woman, Eve, were the two people who acquired knowledge of good and evil, and that event preceded the birth of any children to the couple. However, I am persuaded that the revelation of this event to one of God's Prophets, because it preceded the invention of written language, became garbled in its retelling down through the generations. Since the name Adam also means "man," and the name Eve also means "woman," the story of how men and women were made independent of God's immediate oversight became confused with the names of the first couple, Adam and Eve. So, it was at the garden in the east of Eden that God revealed to man that he was making man independent. From that point on, man would have to cope without benefit of God's immediate and personal help. With few exceptions, God was retreating to a spiritual background from which He would only be accessible through prayer.

In the following way, God's prophet revealed the story of how God made Adam and Eve's clan independent:

Now, the serpent was more crafty than any of the wild animals the Lord God had made. He said to the woman, "Did God really say, 'You must not eat from any tree in the garden'?"

The woman said to the serpent, "We may eat fruit from the trees in the garden, but God did say, 'You must not eat fruit from the tree that is in the middle of the garden, and you must not touch it, or you will die.'"

"You will not surely die," the serpent said to the woman. "For God knows that when you eat of it your eyes will be opened, and you will be like God, knowing good and evil."

When the woman saw that the fruit of the tree was good for food and pleasing to the eye, and also desirable for gaining wisdom, she took some and ate it. She also gave some to her husband, who was with her, and he ate it. Then the eyes of both of them were opened, and they realized they were naked; so they sewed fig leaves together and made coverings for themselves.

Then the man and his wife heard the sound of the Lord God as he was walking in the garden in the cool of the day, and they hid from the Lord God among the trees of the garden. But the Lord God called to the man, "Where are you?"

He answered, "I heard you in the garden, and I was afraid because I was naked; so I hid."

And he said, "Who told you that you were naked? Have you eaten from the tree that I commanded you not to eat from?"

The man said, "The woman you put here with me – she gave me some fruit from the tree, and I ate it."

Then the Lord God said to the woman, "What is this you have done?"

The woman said, "The serpent deceived me, and I ate."

So, the Lord God said to the serpent, "Because you have done this, cursed are you above all the livestock and all the wild animals! You will crawl on your belly and you will eat dust all the days of your life. And I will put enmity between you and the woman, and between your offspring and hers; he will crush your head, and you will strike his heel."

To the woman he said, "I will greatly increase your pains in childbearing; with pain you will give birth to children. Your desire will be for your husband, and he will rule over you."

To Adam he said, "Because you listened to your wife and ate from the tree about which I commanded you, 'You must not eat it,' cursed is the ground because of you; through painful toil you will eat of it all the days of your life. It will produce thorns and thistles for you, and you will eat the plants of the field. By the sweat of your brow you will eat your food until you return to the ground, since from dust you are and to dust you will return."

The Lord God made garments of skin for Adam and his wife and clothed them. And the Lord God said, "The man has now become like one of us, knowing good and evil. He must not be allowed to reach out his hand and take also from the tree of life and eat, and live forever." So, the Lord God banished him from the Garden of Eden to work the ground from which he had been taken. After he drove the man out, he placed on the east side of the Garden of Eden cherubim and a flaming sword flashing back and forth to guard the way to the tree of life.

At this point, God Almighty discarded His *Covenant of Nurture*. In its place. He established a new covenant. I will call this, *God's "Covenant of Tough Love."* Why Tough Love? Because God knew he was intentionally forcing man to cope, on his own, with a multitude of unpredictable hardships over a span of almost 200,000 years from that time to historical times.

Chapter 5. God's Second Covenant: His Covenant of Tough Love — Eden.

There is compelling scientific evidence supporting the dating of man's creation, and locating that event in east Africa. Fairly recently, geneticists have been able to calculate that God's creation of Adam and Eve occurred about 200,000 years ago. That calculation is based on the pattern of variations presently occurring around the world in the structure of human mitochondrial DNA. Mitochondrial DNA exists outside of the cell's nucleus in organelles, from where it directs the energy functions within the cell. Mitochondrial DNA is not part of the main 47-gene DNA that exists in the nucleus of every human cell – what we call "nuclear DNA." Nuclear DNA is the molecular structure that contains the complete genetic code that makes every human a unique, one-of-a-kind individual (except for identical twins). Nuclear DNA is the result of the combination at conception of the DNA structure in the awaiting female egg, and the DNA structure of the successful male sperm. In contrast, mitochondrial DNA is already present in the awaiting female egg, and is unaltered by the male sperm. Therefore, the structure of mitochondrial DNA is passed to offspring only by the mother. Like nuclear DNA, mitochondrial DNA undergoes occasional mutations, and these mutations are passed onto offspring. Mitochondrial DNA's "from-mother-only" feature greatly facilitates a study of patterns in human origin and migration. These studies show that there are relatively few mutational differences among people today, whether they are Africans, Asians, Europeans, Native Americans, or even Pygmies or Australian Aborigines – all of us belong to the same close genetic family. There are two main branches to that family – the Africa branch and the Out-of-Africa branch (which actually includes some people living in Africa). The oldest branch appears to be the Africa branch, for it contains many more mutations than the Out-of-Africa branch. These observations and other studies indicate that God created man at some spot in Africa. And man seems to be a newcomer on earth, showing only one-tenth the mitochondrial variation that is present in our closest living genetic relative, the chimpanzee. This small mitochondrial DNA variation indicates that man's creation occurred about 200,000 years ago.

Perhaps a footnote is appropriate regarding today's intellectual controversy between scientists who draw conclusions from the fossil record, and scientists who draw conclusions from the genetic variation among present-day people.   Defending their advocacy of reliance on genetic patterns among the living, geneticists Allan Wilson and Rebecca Cann wrote in *Scientific American*: "Living genes must have ancestors, whereas dead fossils may not have descendants. Molecular biologists therefore know the genes they are examining must have been passed through lineages that survived to the present; paleontologists cannot be sure that the fossils they examine do not lead down an evolutionary blind alley."

It seems that man spent the first 90,000 years of his 200,000-year existence-to-date living in Africa. There he procreated and learned to defend against wild animals, to gather nuts, fruits and grains, to hunt animals for meat, to maintain interpersonal relations, and, most importantly, to exterminate the animals that most closely resembled him – the hominids. Apes, chimpanzees and monkeys did not compete very much with man, so they consequently escaped his war-like nature. Not so homo erectus and other creatures we call hominids – creatures like the Neanderthal and those in Olduvai Gorge, whose fossil records have been so well publicized by the Leakey's and *National Geographic* magazine. Man was killing them off to the last creature. Only the Neanderthals, the last of the hominids, located further away to the north would remain until their final extinction.

By the end of that first 90,000-year period, man had congregated into villages, and had built temples to God. Instead of ranging far and wide, man was settling down amid the comforts of east Africa. God had commanded Adam and Eve's descendants to procreate and populate the whole earth, but man was still clinging to his original home in Africa. There was still plenty of room in Africa, and it seems man had not yet acquired the explorer's drive. Furthermore, the earth's climate had been rather cold over the first 90,000 years of man's existence. Then, a brief warm period emerged that lasted a few thousand years. One would think that this warm weather alone would have encouraged man to venture out of Africa. But, even then, man did not want to leave Africa. So, at this point, God applied pressure to push man out of Africa. Later, he would reveal that event to his prophets, and they have passed the story down to us. The most dramatic event – the story passed down to us – took place at a town where man was building an unusually tall mud-brick tower in praise of God, and to bring glory to their village, one would suppose. It was called the tower of Babel. They obviously wanted to settle down in that village, but God wanted them to disperse across the Earth. So, God dispersed them by confusing their language and setting them against each other. That push launched some out of Africa, even though most still stayed behind. But, enough left to satisfy God. Man's reluctance to venture into the unknown is understandable. After all, burley and strong hominids lived outside of Africa – the Neanderthals. They lived along the essential out-of-Africa pathway: from the eastern boundary of present-day Egypt, across today's Middle East, northward through today's Turkey and into Europe, and eastward through today's Iran and into Asia. To migrate out of Africa, man had to move into the land of the Neanderthal. But, about 110,000 years ago some did just that.

Now, let's discuss the dispersal of man and the evolution of the races of man. Again, I am presenting my view of this history, but from this point forward, I can advise that it is supported by substantial scientific evidence. Leaving Africa about 110,000 years ago, man first settled in the region we now call the Middle East.

It was in the area of today's Middle East that man had important contact with Neanderthal. Man and Neanderthal were very similar genetically: 99.7 percent of their nuclear DNA structure matched. So man and Neanderthal must have shared some of the same diseases and tendencies toward certain genetic mutations. In fact, recent genetic testing of Neanderthal bones, reported in scientific journals in 2010, suggests that a few Neanderthal-matching genes have been found to have been added to Out-of-Africa-Man's nuclear DNA structure after contact with Neanderthal in the Middle East. These few matches might be the result of a limited gene transfer from Neanderthal to man through sexual union, or be because exposure to the same caves and the same environment encouraged independent mutations of the matching genes, those mutations proving good for Neanderthal and also good for man. On the other hand, the same genetic testing indicates that Neanderthal did not acquire genetic mutations that matched to man. So the matches, few in number, were man matching Neanderthal, not Neanderthal matching man. Approximately one in every fifty genetic variations among Europeans and Asians had a match in Neanderthal. So the genetic influence was small. However, those Neanderthal matches are uniformly present in Europeans and Asians. Yet Africans almost never show evidence of those Neanderthal matches. This is an example of a few of the genetic differences that distinguish Africans from Out-of-Africans (Europeans and Asians).

No matches were found between Neanderthal mitochondrial DNA and Out-of-Africa human mitochondrial DNA. Since mitochondrial DNA passes, unaffected by the father, from mother to child, this lack of any matches suggests that the rape of women by Neanderthal males, if it did produce a viable female baby, resulted in a female baby that was rejected by the mother's clan, or that female baby, if accepted as a girl, was later infertile or was later rejected by human males and did not procreate in significant numbers.

Why did the people who remained in Africa not acquire these few genes mentioned above, which were common to Neanderthals? It was either because there was no sexual union; because they were not exposed to Neanderthal caves or its environment, or because those particular mutations were not good for Africans. So, from the outset of man's migration out of Africa, in the Middle East, there were a few genetic changes that matched Neanderthal, and those changes persist in today's European and today's Asian races.

Although the migration out of Africa was small in numbers, man left in sufficient numbers to eventually populate all of the other continents. Leaving the Middle East, man continued on to the east and even to the north. By 60,000 years ago, man was in India and Asia. By 40,000 years ago man was in Indonesia and Australia. By 35,000 years ago, man had populated Europe, even into the northern reaches of Scotland and Norway. By 30,000 years ago, man had crossed over the Bearing land bridge, and begun the first wave into North and South America. But, this migration was not a continuous and pleasant event, because during 80,000 years of this 110,000-year period, (between 100,000 and 20,000 years ago), the most-northern faction of mankind had become entrapped in a progressively worsening Ice Age. People who had ventured north of the Alps, the Himalayas, and the intervening mountain chains in Europe and Asia, had become trapped between the expanding northern ice cap and the glacier-covered mountains to the south, which their ancestors had crossed over in warmer times. Most of them died from exposure, but a remnant won pitched battles for possession of caves, and, in the caves, survivors stubbornly endured for generation after generation. The most dramatic cave-entrance battles occurred between man and the Neanderthal hominids.

The Neanderthal was naturally better adapted to the cold than was man. It was muscular and heavy-set. It had more hair, though it still needed the caves it had possessed for many generations. Neanderthal had its cave; man needed it. But man could wrestle a Neanderthal about as well as he could wrestle a bear, so one-on-one was not the chosen battle tactic. Although not as strong, man could run faster, a great help when escaping from a bad battle match-up. So, a group of cunning men could entrap one to three Neanderthal males out on the hunt, and eventually severely deplete the number of males living in a cave. Then, they could launch a successful direct attack on the cave and slaughter the females and their young. So, by group effort, cunning tactics and persistence, man gained possession of essential Neanderthal caves. Over thousands of years, the Neanderthal population, the last of the hominids, diminished under intense pressure from man. Man was exterminating the remnant of remaining Neanderthals. The winners in the struggle for the best land were the survivors who had won possession of Ice Age cave habitats. Although perhaps inferior in strength compared to Neanderthals, these survivors were more determined; more cunning; more persistent, and more likely to raise children with those characteristics. By 30,000 years ago Neanderthal was extinct or very nearly extinct, a victim of man's war-like nature and his cunning, group-effort battle tactics.

But, man's population in the cold northern climes was also reduced, particularly among the less determined, less cunning, less persistent individuals. Consistently then, over thousands of years, a remnant of man survived in the caves of the northern climes. The survivors were the ones who were most determined, most cunning, most persistent, and most adept at coping with the harsh environment. Those traits helped them survive to raise successful families. Athletic skills were less important to survival in the cold northern environment than they were in Africa, so athletic skills became somewhat secondary.

Now, dark-skinned beings living in caves lose their pigmentation. That is true of fish, crawfish, crickets, beetles, and salamanders. It is also true of man. I know, because I was an avid cave explorer in my youth. Although the men donned furs and ventured out on hunts, the women largely remained behind in the cave. And the children spent many years almost entirely inside the cave. What little sunlight children received produced more vitamin D within the bodies of the lighter-skinned children, and this contributed to succeeding generations loosing their skin pigmentation. In fact, by 32,000 years ago, long before the last Ice Age cycle abated, light-skinned, cave-dwelling survivors in the upper Danube Valley in Europe had developed the essential textile skills that were so vital to catching fish and small game, and keeping warm. Some were making needles and fish hooks; rope, nets and fishing cord; blankets, cloth, bags, mats and wall coverings. So, when earth's weather began to gradually return to the weather we normally experience today – a process that began 20,000 years ago – white-skinned man emerged from the caves of central Europe and light-yellow-skinned man emerged from the caves of northwest China. These two races of men, although at that time extremely small in population, would procreate with outstanding success and dominate the world.

As the light-skinned cave-dwelling survivors crossed over the mountains, and expanded their territory southward, they forced from the best land those dark-skinned people who, during the Ice Age, had enjoyed the luxury of living in the milder southern climates. Back in Africa, man had remained mostly isolated from the people who had left that continent 110,000 years ago. African man had expanded to south Africa and to west Africa, but still made little effort to move north into Europe or northeast into Asia.

When warm weather returned at the end of the last Ice Age, man, having begun to arrive in North America from northeast Asia about 11,000 years ago, thrived in his new homeland and to the south with greater ease than perhaps anywhere else on Earth. In the Americas, there were no hominid animals like Neanderthal to be killed, and no close relatives like the chimpanzee in whose bodies dangerous viruses and bacteria could adapt for new attacks upon the health of man. In fact, because his journey through northeast Asia and through northwest North America had taken many years, and had involved passage through very cold climate, the prolonged trek had "acted as a disease filter," so that "most of the illnesses which afflicted men in the Old World were left behind." Game was plentiful, and animals were not accustomed to defending themselves against man. So, man experienced a huge population explosion as he advanced southward along the Pacific coast, through Central America, and on into South America. And, he moved eastward across mountains and deserts to arrive in eastern North America, where he likewise thrived. Furthermore, he crossed the jungles to reach southeastern South America, and there, too, he found a healthy climate. And, large game animals fell before man's advance, notable species being killed to extinction. Victims included the mammoth, the horse, the camel, the ground sloth, and a relative of the American bison. By 9,000 years ago, these were all extinct. Yet, the American bison, elk, moose and deer remained, deer being the most useful game animal in southeastern North America. Once man's expansion from the Pacific Northwest began in earnest, he perhaps completed his occupation of North and South America within one thousand years, for he found little to impede his advance. Man seemed to be present throughout the Americas by 9,000 years ago. But, as each generation was born, less and less resistance to Old World diseases remained. So, when, 8,500 years later, Native American people suddenly, and desperately, needed for their bodies to fight those diseases, their biological defenses remembered not how. So, beginning in 1500 A. D. Native American people would die by the millions as Spanish invaders, based in the Caribbean and the Gulf of Mexico, unknowingly dispensed upon a helpless people wave after wave of small pox, measles, typhus, tuberculosis, chicken pox, and influenza.

That concludes the history of how man spread across the Earth and, in the process, diversified into races, and we understand that this human expansion was an evolutionary process. Now, let's return to my thoughts on the role that God must have played in this great expansion, this great migration. Since, at about 10,000 years ago, man had spread throughout the earth and eliminated all hominid animals, this was most probably the time that God decided to make a new covenant with man.

Perhaps, I should recap "Understanding Creation and Evolution" up to this point:

About 200,000 years ago, God created man and nurtured the immediate descendants of Adam and Eve, to protect them, and to then teach them survival skills. Then, perhaps 1,000 years later, He backed off and let man fend for himself, wrestling with good and evil as best he could. Then, at about 110,000 years ago, God felt the need to push man out of Africa, for, in spite of the fact that the weather to the north had become warm, man seemed far too content to stay put in environmentally friendly eastern Africa. So, during this protracted warm weather cycle a significant number did move out of Africa, and man populated much of India, Asia and Europe. But later, a progressively worsening ice age trapped the people that had been lured into the northern climes by unusually warm weather. Those trapped in the cold north, who were lucky enough to conquer cave sites, took refuge within, and that remnant survived to multiply and become the two most successful races among the races of man, the result of natural evolution.

Chapter 6: God's Third Covenant: His Covenant of Adoption — Noah.

The time was approaching when God would advance to a new covenant. which I will call, His *Covenant of Adoption.* To understand this event, you need to know some meaningful background information. So, I began this segment with the story of Black Lake.

During ice age times, white-skinned man was attracted to the edge of Black Lake, a somewhat salty, land-bound lake centered on today's Black Sea. Back then, Black Lake lay within a huge basin that sloped down to the water, whose surface was then about 500 feet below sea level. Being at such a low altitude, the land around Black Lake was relatively warm, considering its latitude. But, man had to continually retreat, because Black Lake was rapidly filling with glacier melt. It eventually rose several hundred feet above sea level before it breached the headwaters of the Sakarya River and began overflowing into the Mediterranean Sea. This flushed out all of the salty water. By 15,000 years ago, the great northern ice cap had retreated so far north that alternate pathways developed to send the melt water to the North Sea. Furthermore, about 12,500 years ago, the weather became colder and much dryer. As a result of these two changes, the water flowing into Black Lake fell below the rate at which water was evaporating from the surface – the overflow into the Mediterranean Sea at Sakarya River stopped, and the lake surface began to fall. But, the lake water had been so thoroughly flushed free of salt that the concentrating effect of the evaporation did not even make the lake water brackish. There was another cold, dry spell 8,200 years ago, but it moderated after 400 years. Two hundred years further toward present time brings us to the days of Noah, a man in whom God was especially pleased. Noah lived at the edge of Black Lake 7,600 years ago.

The people around Black Lake in the days of Noah were accomplished farmers and fishermen. They cultivated wheat and barley and herded domesticated goats. They built boats and fished the lake. They built houses of differing designs. They traded between villages. They may have even smelted copper, because 6,500-year-old evidence of that technology can be found today from the nearby Balkans southward to Israel.

In fact, the Black Lake people probably represented the most advanced civilization on earth at that time. And it appears that, in the days of Noah, 7,600 years ago, God was ready to reveal Himself more directly to man. The time for harvesting souls was arriving. God wanted to make a covenant with a family, so that, through them and their descendants, His presence would be revealed to others. He wanted to make a new and special covenant. For this purpose He chose Noah, for among all of the families on Earth, Noah's stood out as the most righteous and the most capable of becoming God's prophet to future generations. And, Noah lived in the most advanced civilization on Earth – the shore of Black Lake, where a dramatic event was about to occur.

In the days of Noah, the Sea of Marmara, which was connected to the Mediterranean Sea by the Dardanelle Straight, had, for about 12,500 years, been rising along with all of Earth's oceans in tandem with the receding ice caps. And, the salt water within the Sea of Marmara was only prevented from plunging down into Black Lake by the earthen hills through which the narrow Bosporus Straight now cuts. Those low hills, which had no underpinning of bedrock, lay atop the eastward-creeping Eurasian Plate, near its fault-line junction with the westward-creeping Anatolian Plate. God knew that, one day, an earthquake would rip those low hills apart, creating a channel through which ocean water could rapidly flood Black Lake. Once the route would become established, erosion would expand the flow hourly. Within days, the flow would be gushing violently. The flow would soon equal that of 200 Niagara Falls. Black Lake would rise six inches every day. Some people living along the shore would see the danger and steadily moved back. But many would figure the rising would stop, and would hang to nearby high ground too long, only to find their escape route cut off. Black Lake valley would be flooded within a few months. God planned to use this event to destroy the most advanced civilization of the day; destroy their homes, farms and way of life; turn their beautiful fresh-water lake into a salt-water sea of death and destruction; force them onto the barren hills beyond; scatter them far and wide. God planned to use that event to separate Noah and his family from all of their neighbors, and from the other clans in the valley. This event would be the Great Flood, which is recorded in several ancient oral histories and in *The Bible*.

So, God came to Noah and told him to prepare for a great flood by building a huge boat according to the set of construction plans, which He laid out before the surely dumbfounded man. How could there be such an enormous flood, Noah must have asked? But, God assured him that it would take place, and the huge boat was essential to his survival and his immediate family's survival. Asking for Noah's obedience, without benefit of physical evidence of compelling danger, was God's way of testing Noah's faith. While neighbors laughed at the hard-working Noah, he and his sons struggled to gather the building materials and construct the huge craft. God could have saved Noah and his family with a much smaller craft, but He wanted Noah and his sons to work especially hard at building a really huge craft. It also seems that God wanted to impress man with his upcoming responsibility, as caretaker of life on earth, for the boat was to hold many animals, some of them quite large. God knew that man's upcoming population explosion would threaten the viability of plants and animals all over the world. Noah's experience was to be a lesson to man: Noah would be the example of a loving caretaker of earth's plants and animals, just as God had always been the loving caretaker of man.

From the Bible we read:

Noah did everything, just as God commanded him.

The Lord then said to Noah, "Go into the ark, you and your whole family, because I have found you righteous in this generation. Take with you seven of every kind of clean animal, a male and its mate, and two of every kind of unclean animal, a male and its mate, and also seven of every kind of bird, male and female, to keep their various kinds alive throughout the earth. Seven days from now, I will send rain on the earth for forty days and forty nights, and I will wipe from the face of the earth every living creature I have made."

After the earthquake produced the fracture, through which seawater broke through the Bosporus hills, Black Lake Valley began to rapidly flood. Although Noah experienced considerable rain, the lake was rising so fast that Noah figured most of the water had to be coming up from springs beneath the ground. *The Bible* account says: *"on that day all the springs of the great deep burst forth, and the floodgates of the heavens were opened."* But, Noah was prepared for the great flood. His boat was complete, and God had helped with gathering up animals. Noah, his immediate family, the animals, and food for all, were safely aboard Noah's huge boat by the time the water rose up around it. All passengers would stay aboard throughout the remaining filling of the Black Sea. Eventually, the flooding stopped. *The Bible* says: *"Now the springs of the deep and the floodgates of the heavens had been closed, and the rain had stopped falling from the sky."* Finally, the huge boat drifted to dry ground in the vicinity of Mount Ararat. There, Noah, his family and the animals disembarked. Before long, Noah must have encountered people and animals that he surely realized had not been aboard the huge boat, but that realization was lost in the retelling of Noah's story down through the generations. As the retold story survived, and was written down to eventually be printed in today's Bible, the Great Flood was said to have killed all land animals and all people throughout earth, save those in Noah's huge boat. Many were surely killed by the flood, including thousands of people, and the environmental destruction rent upon the land was obviously complete and total, but, by the time of the Great Flood, man had migrated around the globe. The story of the death by drowning of all the people and animals trapped around Black Lake became the story of the alleged death by drowning of all the people and animals living all over the earth. That expansion of scope is easy to understand given that, in those days, a person's knowledge of geography was limited to his immediate region.

Because of his faithfulness prior to the Great Flood, God made a covenant with Noah. Through him and his descendants, God pledged to shepherd a great people, and those people would be the conduits through which God would reveal Himself to man throughout the world. Noah's descendants would become the Jewish people, and Israel would eventually become their homeland.

God's prophet reveals to us the story of how God made his covenant with Noah and his family:

> *Then, Noah built an altar to the Lord and, taking some of all the clean animals and clean birds, he sacrificed burnt offerings on it. The Lord smelled the pleasing aroma and said in his heart: "Never again will I curse the ground because of man, even though every inclination of his heart is evil from childhood. And never again will I destroy all living creatures, as I have done. . .*

> *"Whoever sheds the blood of man, by man shall his blood be shed; for in the image of God has God made man.*

As for you, be fruitful and increase in number; multiply on the earth and increase upon it."

Then, God said to Noah and to his sons with him: "I now establish my covenant with you and with your descendants after you and with every living creature that was with you – the birds, the livestock and all the wild animals, all those that come out of the ark with you – every living creature on earth. I establish my covenant with you: Never again will all life be cut off by the waters of a flood; never again will there be a flood to destroy the earth."

Survivors of the destruction of the Black Lake region would migrate to other regions, taking with them the skills they had developed in what had been man's most advanced civilization. Farmers called Vinca would move into the plains of present-day Bulgaria. Others would migrate into present-day Greece. People called Linearbandkeramik would flee up the Dniester River, and on across Europe as far as present-day Paris. Others would move up the Dnieper, the Don and the Volga into Russia. Others would flee east to the Caspian Sea and domesticate the horse, with which they would master the huge expanse of land to the east. A group, perhaps looking for a place to apply their irrigation farming skills, would migrate south past the Mediterranean and into the Nile Valley, where they would lead in the expansion of Egyptian civilization. But many, including Noah's family, would travel south-southeast across the mountains into the Tigris and Euphrates Valley, where they would establish irrigation-based farming where practical. Eventually, Noah's descendants would move further downstream to the delta region and contribute to the great Sumerian Civilization.

That is the story of Noah, which most of you will have heard before, but probably not in the manner that I have just presented it. By the way, there is a growing body of new and exciting scientific evidence to support my assertion that Noah's flood occurred at Black Lake.

God's covenant with Noah was His third Covenant, which I am naming, God's *"Covenant of Adoption."* Why? Because, with this covenant God signaled that He was adopting the Jewish people, so that He would be revealed to man through them. This was the first of three covenants that God would make with His Adopted People, the Jews. Later, He would make a covenant with Abraham, and then He would make another with Moses. As mentioned earlier, Noah's people would cross the mountains of present-day Turkey, and make their way to the Euphrates delta. In that land, 2,500 years later, God would speak to Abram.

Yes, about 4,100 years ago, God made another covenant with man. He made a covenant with Abram, who he renamed Abraham. Abram was a descendant of Noah. This was God's fourth covenant, which I shall call, God's *"Covenant of Expanding Presence."*

Chapter 7: God's Fourth Covenant: His Covenant of Expanding Presence — Abraham to Moses.

At the time God initiated His *Covenant of Expanding Presence*, Abram lived in the city of Ur, in the land of the Chaldeans (in present-day southern Iraq). God told Abram to leave the land of his people and migrate to Canaan (in present-day Israel) – a new land that was destined to become the land of God's people. Abram obeyed God and moved to the land of Canaan. There, God gave Abram land. Shortly thereafter, God revealed His wrath toward sinful people, for, before Abram's eyes, the people of Sodom and Gomorrah were destroyed because of their blatant sinfulness.

God's prophet reveals to us the story of how God initiated a covenant with Abraham and his descendants:

> The Lord had said to Abram, "Leave your country, your people and your father's household and go to the land I will show you. I will make you into a great nation and I will bless you; I will make your name great, and you will be a blessing. I will bless those who bless you, and whoever curses you I will curse; and all peoples on earth will be blessed through you." . . .

> "On that day the Lord made a covenant with Abram, and said, "To your descendants I give this land, from the river of Egypt to the great river, the Euphrates – the land of the Kenites, Kenizzites, Kadmonites, Hittites, Perizzites, Rephaites, Amorites, Canaanites, Girgashites and Jebusites."

God's prophet reveals to us the story of how God intervened in the natural course of life, so that Abraham and his wife Sarah, far older than normal childbearing ages, could conceive and give birth to a healthy baby, whom they named Isaac:

> When Abram was ninety-nine years old, the Lord appeared to him and said, "I am God Almighty; walk before me and be blameless. I will confirm my covenant with you and greatly increase your numbers." You will be the father of many nations. No longer will you be called Abram; your name will be Abraham. "As for Sarai your wife, you are no longer to call her Sarai; her name will be Sarah. I will bless her, and will surely give you a son by her. I will bless her so that she will be the mother of nations; kings of peoples will come from her." Abraham fell face-down; he laughed and said to himself, "Will a son be born to a man a hundred years old? Will Sarah bear a child at the age of ninety?" And Abraham, concerned over the fate of his son Ishmael and Ishmael's people, said to God, "If only Ishmael might live under your blessings!" Then God said, "Yes, but your wife Sarah will bear you a son, and you will call him Isaac. I will establish my covenant with him as an everlasting covenant for his descendants after him. And, as for Ishmael, I have heard you: I will surely bless him; I will make him fruitful, and will greatly increase his numbers."

God later continued his covenant through Isaac and Isaac's descendant, Jacob. At one encounter, God told Jacob that he was to take the new name "Israel." At another encounter, God assured Jacob of the continuing covenant:

> *"And God said to [Jacob], 'I am God Almighty; be fruitful and increase in number. A nation and a community of nations will come from you, and kings will come from your body. The land I gave to Abraham and Isaac I also give to you, and I will give this land to your descendants after you.'"*

God also continued his covenant through Joseph, a descendant of Jacob, the man God had renamed "Israel." Joseph, while a young man, was sold into slavery by some evil brothers. This occurred about 3,900 years ago. The buyer was a group of traveling foreign merchants, who took him to Egypt, in northeast Africa, and resold him to Potiphar, an official in the Pharaoh's government. Considering that Joseph was a simple herder of goats and sheep, the resulting exposure to urban, highly organized, and structured Egypt must have been a powerful educational experience. Initially prospering as a bonded man, Joseph was given full responsibility for overseeing Potiphar's house and field servants. But, after some time, Potiphar's wife became vicious toward Joseph, and had him thrown into prison. But, God blessed Joseph while in prison, and permitted him to eventually impress the Pharaoh himself by his ability to correctly interpret dreams. In fact, the superstitious Pharaoh was so impressed with Joseph's predictive power that he told Joseph, *"I hereby put you in charge of the whole land of Egypt."* So, Joseph became the Pharaoh's chief administrator. And, it was fortunate for the Israelites that Joseph was in a position to help them, for, about this time a severe famine in their land forced all of them to migrate, along with their herds and flocks, to Egypt, in search of pasture and food. The famine lasted several years. Leaders among the newly arrived Jews explained their plight before Pharaoh and Joseph: *"Your servants are shepherds, just as our fathers were. We have come to live here awhile, because the famine is severe in Canaan and your servants' flocks have no pasture. So now, please let your servants settle in Goshen."* In response, Pharaoh instructed Joseph: *"Let them live in Goshen. And if you know of any among them with special ability, put them in charge of my own livestock."*

Now, with God's help, Joseph had anticipated the famine and, as chief administrator, had arranged for, in previous fruitful years, the storing up of huge grain reserves. Joseph made sure the Israelites received food, but, in return, they eventually lost possession of their livestock and became bonded to the Egyptian government, as did the other peoples in the region.

Long after Joseph's death, a new Pharaoh came to power, and became fearful that the now-numerous Jews might become politically powerful: *"The Israelites have become much too numerous for us. Come, we must deal shrewdly with them, or they will become even more numerous, and, if war breaks out, will join our enemies, fight against us and leave the country."* So, Pharaoh *"put slave masters over them to oppress them with forced labor, and they built Pithom and Rameses as store cities for Pharaoh. But, the more they were oppressed, the more they multiplied and spread; so the Egyptians came to dread the Israelites and worked them ruthlessly. They made their lives bitter with hard labor in brick and mortar, and with all kinds of work in the fields.*"

Yet, in spite of this oppression, these simple Israeli herdsmen had learned much from their owners that future generations would use to advantage. And exposure to this learning experience seems to have been the reason God had permitted his chosen people, for a few generations, to become enslaved in Egypt. In fact, the Jewish experience resembled that of those West Africans who would, much later, be sold into slavery by their fellowmen – their African neighbors – transported to North and South America by foreign merchants, and resold to farmers in that distant land. They, too, were destined to learn much from their owners – much that future generations would use to advantage.

Long after Joseph's death, and after several generations of oppression at the hands of Pharaoh's government, God chose a man to lead His people out of bondage. That man was Moses. One day, while Moses was tending a flock, God presented himself from within a burning bush: "*I am the God of your father, the God of Abraham, the God of Isaac and the God of Jacob. I have indeed seen the misery of my people in Egypt. I have heard them crying out because of their slave drivers, and I am concerned about their suffering. So, I have come down to rescue them from the land of the Egyptians and to bring them up out of that land into a good and spacious land. . . . I am sending you to Pharaoh to bring my people the Israelites out of Egypt.*" And, Moses did as God commanded, and presented God's demand to Pharaoh. A contest ensued between God's miraculous signs, as dispensed by Moses and his assistant, Aaron, and the competing magic tricks dispensed by Pharaoh's court magicians. The miraculous signs became deadly, as God, through the hand of Moses, put the Egyptians through eleven plagues: blood, frogs, gnats, flies, livestock, boils, hail, locusts, darkness, and the firstborn. After the plague of the firstborn, the terrified Pharaoh summoned Moses and Aaron and pleaded: "*Up! Leave my people, you and the Israelites! Go, worship the Lord as you have requested. Take your flocks and herds, as you have said, and go. And also bless me.*" This took place about 3,450 years ago. This was God's fifth covenant, which I shall refer to as, God's "*Covenant of National Presence.*"

Chapter 8. God's Fifth Covenant: His Covenant of National Presence — Israel.

The Israelites quickly gathered together and departed Egypt. *"By day, the Lord went ahead of them in a pillar of cloud to guide them on their way, and, by night, in a pillar of fire to give them light, so that they could travel by day or night."* Before long, Pharaoh's heart hardened and he dispatched chariots and horsemen to give pursuit. This military force threatened to trap the Israelites against the coast of the Red Sea, but God created an immense wind that swept a dry corridor across the seabed, and the Israelites safely crossed through. Soon thereafter, Pharaoh's chariots and horsemen entered the seabed corridor in hot pursuit. However, they all drowned when God permitted the wind to stop so the sea would flow back upon them. The Israelites were thereby safely out of Pharaoh's reach.

For about fifty years, God led Moses and the Israelites, along with their sheep and goats, along a wandering route between Egypt and Canaan, much of it in the desert. During this time, God often appeared through His cloud. A mobile Tent of Meeting was constructed according to God's design, so they could worship in God's House wherever they roamed. And the primary religious treasures were kept in a mobile "Ark of the Covenant" for ready transport. Israeli religious ceremony became highly structured. Numerous laws, inspired by God, structured Israeli society and individual living practices. At one point, God carved upon two stone tablets the laws we call the "Ten Commandments," and directed Moses to present them to the Israelites. The tablet began with *"I am the Lord your God, who brought you out of Egypt, out of the land of slavery."* The Ten Commandments demanded:

1. *"You shall have no other Gods before me. . .*
2. *"You shall not make for yourself an idol . . .*
3. *"You shall not misuse the name of the Lord your God . . .*
4. *"Observe the Sabbath day by keeping it holy . . .*
5. *"Honor your father and your mother . . .*
6. *"You shall not murder. . .*
7. *"You shall not commit adultery. . .*
8. *"You shall not steal. . .*
9. *"You shall not give false testimony against your neighbor. . .*
10. *"You shall not covet . . ."*

After fifty years of purposeful wandering in the desert regions between Egypt and Canaan, God apparently determined that His people were ready to invade the land of Canaan and settle down upon the *"Promised Land."* The wandering phase of their experience was complete. At this point, Moses died, his mission concluded. Joshua assumed leadership of the Israeli people. Under Joshua, the Israelites invaded Canaan by first taking the fortified city of Jericho. Although the walls of Jericho quickly fell with God's help, the fighting to conquer Canaan, and to build a unified nation, was a long process, consuming many generations.

Eventually, a great king, named Saul, arose, and Saul's reign marked the start of the final stage of a movement to expand and unify God's chosen people as the nation of

Israel. Saul was born about 3,050 years ago. David succeeded Saul. David led his people in the conquest of the land that would become Jerusalem and much surrounding territory, defeating the Ammonites and the Philistines, and making Israel a prominent nation. Solomon succeeded David. Solomon oversaw the construction of the first permanent Temple at Jerusalem: *"In the four hundred and eightieth year after the Israelites had come out of Egypt, in the fourth year of Solomon's reign over Israel, in the month of Ziv, the second month, [Solomon] began to build the temple of the Lord."* That was about 2,950 years ago.

About 360 years after Solomon built the temple in Jerusalem, Nebuchadnezzar led the Babylonian army in the conquest of the Israeli people and their capital at Jerusalem. The Babylonians destroyed the Temple, and, by 586 BC, had carried the more prominent Israeli people back to Babylonia to serve the Babylonian Empire. Eventually, the Persian Empire absorbed the Babylonians, and the Israeli exiles came under the control of Cyrus, King of Persia. Wishing to help the dislocated Israeli people, Cyrus proclaimed: *"The Lord, the God of Heaven, has given me all the kingdoms of the earth, and he has appointed me to build a temple for him at Jerusalem in Judah. Anyone of his people among you – may the Lord his God be with him, and let him go up."* And in time, under the leadership of Ezra (458 B. C.) and Nehemiah (432 B. C.), the Israeli people returned to their homeland and rebuilt Jerusalem and their temple to God. It seems the Israeli people were learning to remain a distinct people capable of surviving the turbulent power shifts that plagued the Middle East. A significant number were not being absorbed into the society of their conquerors, as were other people in the region. God intended that the Israelis not be assimilated, but survive as a people, for they were God's messengers to all mankind. This concludes the history, up to the emergence of the Christian era, as my study as led me to understand it. I hope the telling of this story, which is surely familiar to many, in the above manner, has sharpened your understanding. We move forward now to the present covenant, presented to man by Jesus Christ.

Chapter 9. God's Sixth Covenant: His New Covenant of Salvation through Faith in Jesus Christ.

The covenant between God and the Jewish people came to full flower about 2,000 years ago, (in either 6 or 5 B. C.) with the birth of Jesus Christ, God's Son of Man. (Perhaps God had, on other planets in the Universe, created intelligent beings with souls, but He wisely refrained from revealing that to man.) So, we are concerned only with God's Son for earth, his Son of Man, Jesus Christ, who took on a human body to walk and teach among the Israeli people.

At the time of the arrival of Jesus, the Roman Empire was extensive and it included the Israeli people. Augustus was Emperor. The Empire's existence, its language and or organization greatly facilitated communication across far-reaching lands. So, God had chosen the era of the Roman Empire to bring his Son to Earth, and there present His *New Covenant* to all mankind. This covenant between God and all mankind – through His Son, Jesus Christ – is the covenant under which we live today. To the people He encountered, Jesus taught much more than had previously been known about the nature of God and his expectations for man. And, through His disciples, His message, or Gospel, was spread throughout the region. Through later generations, His message has been spread throughout the world. His disciple, John, summarized the New Covenant as follows:

> *"For God so loved the world that he gave his one and only Son, that whoever believes in him shall not perish, but have eternal life. For God did not send his Son into the world to condemn the world, but to save the world through him. Whoever believes in him is not condemned, but whoever does not believe stands condemned already, because he had not believed in the name of God's one and only Son."*

So, God's covenant with the Jewish people was supplemented by the New Covenant with all mankind through Jesus Christ. This was God's sixth covenant, which I shall call, God's *"Covenant of Salvation Through Faith in Christ."* Today, all people on earth live under *God's Covenant of Salvation Through Faith in Christ.* Yet, subsequent history suggests that a residual covenant must still exist between God and the Jewish people, although the teachings of Jesus would indicate otherwise.

By the time Europeans began to settle the Americas, Christianity was the only significant religion in their European homeland. The original Christian Church was the Roman Catholic Church, and it had split into an eastern and a western branch long before settlement of the Americas began. The two branches of the Roman Catholic Church built elaborate, extensively decorated stone cathedrals throughout Europe, elevated Mary to a companion god, and interposed its priests between God and man. Although no lay Europeans spoke Latin in everyday life, the Roman Catholic Church continued to present its teaching and rituals in the old Latin language. Close ties between the Roman Catholic Church and European governments ensured both stayed in power, and received ample revenue from the people. But, the invention of the printing press, and the preaching of Protestant activists, such as Martin Luther (German, 1483-

1546), John Knox (Scotsman, 1514-1572) and John Calvin (Frenchman, 1509-1564), ushered in movements throughout the British Isles and Europe aimed at placing translated Bibles in the hands of the people. Europeans of modest education were able to read *The Bible* in their native language and acquire a personal relationship with God – without the services of a priest. The original Gospel text revealed to European readers a clear image of God and Jesus, and caused many to criticize the official teachings of the Roman Catholic Church, because they realized that *The Bible* did not validate the deity of Mary, and did not authorize priests to act as intermediaries between man and God. Many left the Roman Catholic Church and formed Protestant churches, dedicated to the basic teachings of Jesus, as revealed in *The Bible*. Instead of priests, these new Protestant churches had pastors and part-time lay leaders. Similar churches within a region were often loosely organized, but they had no "pope." Very few Protestant churches had ties with national governments, one notable exception being the Church of England.

Chapter 10. God's New Covenant Arrives in North and South America.

Europeans discovered the Caribbean Islands and the American continents beginning in 1492. The Spanish arrived first, bringing their Catholic religion and priests. English settlement of North America began at Jamestown, Virginia Colony in 1607, bringing the mainstream English Protestant religion. Jamestown was thriving by the time Puritan Separatists arrived at Massachusetts Bay in 1620 to begin the founding of a Puritan-based colony east of the Dutch colony of New Netherland, which had been founded in 1614. New Netherland would become New York and the area settled by Puritan Separatists would become New England. Roman Catholics, the dominant religion in Spanish and Portuguese-settled Central and South America, were a small minority in the settlement of North America, and remains so to this day. In North America several Protestant churches rose to prominence. The Baptist, Methodist, Presbyterian, Episcopal and Lutheran Churches were the most important in Colonial America, especially in the southern colonies and subsequent states. The Society of Friends was important in Pennsylvania.

Additional understanding of the religious issues in what would become New England is worthy of mention:

Soon after the settlement of what would become known as New England, the Puritan Faction of the Church of England became increasingly important back home in the British Isles, but created political enemies there by striving to "purify" the souls of non-members through political action. Discouraged by that rejection, many more Puritans, called Puritan Separatists, left England for Massachusetts, and played a major role in further settling what would become known as New England. The Unitarian church later arose, but its membership did not fully embrace the teachings of Jesus Christ. A faction within that Church, called Transcendentalists, adopted a creative approach to finding God and establishing so-called moral truths. Their mix of non-Christian religious beliefs permeated New England literary circles, and, between 1830 and 1875, played a major role in establishing a northern States culture of righteous indignation toward Southern States people, which escalated into the War Between The States.

Southern States people, both those of European ancestry and those of African ancestry, mostly embraced the religious beliefs expressed by Baptists, Methodists, Episcopalians and Presbyterians. They had little to do with Unitarians or Transcendentalists.

Chapter 11. God's New Covenant and Africans in North and South America.

Early in the presentation of my Understanding of Creation and Evolution, I wrote about the migration of a faction of man from Africa northward to populate Europe and Asia and of the horrific hardships they suffered during the Great Ice Age. And I wrote of the faction of mankind that stayed behind in the African homeland and, as luck would have it, avoided the challenges of surviving harsh Ice Age weather. We now arrive at a time when the African and the European are brought together in large numbers in the Caribbean and North and South America. These Africans arrived as slaves, bonded persons purchased by the European settlers. Today people of African ancestry (from pure to slight African ancestry), as independent individuals, live among people of European ancestry all as citizens of their respective countries. The record of this second great "Out of Africa" experience needs to also be understood.

Let us look at the numbers. During the colonial days of the Caribbean Islands, South America, and North America, how many people from sub-Sahara Africa were enslaved, sold and shipped across the Atlantic Ocean? I will give you the best available answer to that question. But first, remember that it was Africans, of neighboring tribes, that captured and enslaved individuals. Europeans did not have to go inland from the South Atlantic Ocean to do the capturing. The captors just bought their victims to various seaports along the coast of Africa. The supply of captives must have overwhelmed the carrying capacity of the merchant sailing ships, because it is estimated that 23,500,000 were enslaved by fellow Africans, but only 13,400,000 were sold to seamen at the West Africa seaports. The difference, 10,100,000, were, for the most part killed or died of disease or malnutrition. It is estimated that 12,000,000 survived the Atlantic crossing and 11,300,000 were successfully sold to owners in the New World. Of the number sold, 7,000,000 are believed to have survived 3 or 4 more years. I have no survival figures beyond the 4 year span.

Now, let us look at how many of those 23,500,000 enslaved Africans were sold to owners in the English colonies of North America. The best estimate is 575,000 sold through the legal cutoff date of 1808, with another 25,000 smuggled in and sold after that. So, only one in forty enslaved Africans were lucky enough to be sold in the English colonies. How does this experience compare to Native Americans? Well, we know that far fewer Native Americans are living today than were living in North America at the time of Columbus. Disease and war almost totally destroyed that vast population. What about those 600,000 Africans that were sold into the English colonies of North America and the states that followed? Has their population grown over time or diminished? Well, it has grown by leaps and bounds! The 2012 United States Census Bureau reported an estimated population of 44,456,000 Americans of pure or partial African ancestry living in America. That is a population that is 74 times the number imported as slaves. That is approaching twice the estimated number of all the Africans enslaved by their fellow men in Africa during the era of the Atlantic slave trade. It is more than three times the number of Africans loaded onto ships at West African seaports for transportation to all destinations in the Caribbean, Central America, South

America and North America. I find it helpful to consider these numbers when confronting sensitivities to racial issues in today's world. Perhaps you will refer to them on occasion as well.

Now, let's fast-forward in time, and explore what life is like in sub-Sahara Africa today. Mostly gone are the economic and governance benefits that had been given to that vast continent during its colonial era, through generations of hands-on involvement from the many, many Europeans who had come to settle there, raise families there, and contribute to each colonial nation important benefits in many fields, including agriculture, manufacturing, finance, and the workings of a civil society. But, the departure, over recent decades, of so many residents of European ancestry and the rapidly growing native African population has made life in Africa today noticeably more difficult for native Africans. I cite Uganda and the genocide between the Hutu and the Tutsi. I cite the economic distress, ethnic terrorism, and criminal lawlessness that have been so prevalent across the continent. Even the showcase nation of South Africa is hurting. I cite the destructive nature of disease and malnutrition. I remind you that Haiti, although not far from Florida, is essentially a nation populated solely by people from sub-Sahara Africa, and, one would suppose, for similar reasons also suffers. So, why do the natives of sub-Sahara Africa suffer without close association, within their societies, with residents of European ancestry? The answer seems to me to be straightforward. The answer is quite evident in a simple reading of history. Here is the answer: "a society made up solely of the native African race will perpetually suffer because it lacks the economic and civil leadership that history tells us can only be generated when a significant part of the population is people of the European or Asian races, and where those races are permitted important political rights. Nations populated solely by Asian people can thrive. Nations populated solely by European people can thrive. Nations populated solely by African people cannot thrive. Nations of mixed races can thrive, if the African race is not dominant.

Chapter 12. Characteristics of the Races of Man.

There are differences in people that derive from their racial ancestry. The differences in appearance are perhaps most obvious, but the differences in various abilities are of more importance. Now, I need not treat the appearance differences among the races of man. I assume all readers are familiar with appearance differences. So, what I need to do is tell you about four non-appearance characteristics of individuals, and describe how those characteristics are distributed among individuals, and among the races of man. These four characteristics are: cognitive ability, athletic ability, musical ability, and traits of character. But first, let me lay a foundation for this discussion by summarizing the basic science of genetic inheritance.

An individual's innate cognitive ability, athletic ability, musical ability, and traits of character seem to result from mixtures of that individual's genetic inheritance and that individual's environmental interactions. The two most important environmental factors are parental care and diet history, from conception through childhood. But, it is important that you understand that environmental factors are less important than genetic factors. Scientific studies, particularly those based on identical twins separated at birth, indicate that genetic inheritance accounts for between 80% and 40% of an individual's innate cognitive ability. I am persuaded to embrace Herrnstein and Murray's approximation of 60% genetic inheritance of cognitive ability. Furthermore, I am persuaded to attribute about 60% of an individual's athletic ability, musical ability, and traits of character to genetic inheritance, as well.

Siblings in a family vary in cognitive ability, athletic ability, musical ability, and traits of character because, at the time each individual is conceived, he or she consists of a unique combination of genetic instructions, which were synthesized from the genetic offerings that were derived from four grandparents. The same genetic mechanism is obviously present when examining differences within, and between, the races of man. If four grandparents of the same race can contribute a genetic mix that results in the span of cognitive ability typically observed among siblings in a single family, then surely it is likely that genetic factors can explain the differences observed between individuals from many families within the same race. Likewise, genetic factors contribute to the differences observed between groups of individuals belonging to different races. Looking at the issue from a different perspective, I am also persuaded that there is no scientific reason to suppose that racial differences are limited to outward appearance factors such a skin and hair color. There is nothing in the workings of genetic reproduction that restricts genetic variations to outward appearance characteristics.

Now, we know that man began the long process of evolving into the African, European and Asian races about 100,000 years ago, and that time span equates to 5,000 generations. It is further important to recognize that there is no known mechanism in genetics that would ensure, over a span of 5,000 generations, that race A and race B would contain the same genetic mix of cognitive abilities, the same average, the same range. So, surely it is likely that, for genetic reasons alone, the cognitive ability, athletic ability, musical ability, and traits of character of millions of individuals in race A can, and ought to, differ from the cognitive ability, athletic ability, musical ability, and traits of character of millions of individuals in race B in both characteristic averages and in characteristic ranges. I will start with how Cognitive Ability varies between people of the same race.

Chapter 13. Distribution of Cognitive Ability among Individuals and between Races.

Innate cognitive ability has probably not changed significantly since Europeans began settling in North America 400 years ago. That was only 20 generations back in time, and genetic change would have been rather minor in only 20 generations, for that is only two fifths of one percent of the 5,000 generation span since man began dividing into races. So, it is possible to look at innate cognitive ability of recent generations, and, from that data, infer the innate cognitive ability of ancestors four hundred years previously. This pattern should also be true for athletic ability, musical ability, and traits of character.

The most accurate measure of cognitive ability appears to be standardized intelligence testing. The most commonly applied test of this type is the well-known intelligence test that is normalized to a score of 100. I have drawn the intelligence test numbers presented in the following paragraphs from *The Bell Curve* by Herrnstein and Murray. I will call this testing, "cognitive testing."

Cognitive testing of European Americans during the twentieth century shows an average score of 103.6, with a median of 103.4. And, 80% of European Americans fall within a range from 86.1 to 121.7.

Similar testing of African Americans shows an average of 87.5. However, the racial ancestry of the African Americans in the test group varied widely and probably averaged about 75% African and 25% European. Therefore, I have subtracted 25% of the difference in European-American and African-American scores to determine a residual African score, more accurately characterized as a sub-Sahara, western Africa score. This I did throughout the distribution of scores. The residual distribution (presumably pure African) shows an average score of 82.1. This distribution indicates that Africans imported into America had an average innate intelligence equivalent to a score of 82.1, with a mean of 83.4. The corresponding distribution of cognitive scores reveals that 80% of these arriving Africans fell within a range from about 72 to about 93. However, twentieth century intelligence testing in sub-Sahara Africa, although limited and hard to analyze, indicates an average score of about 75 for people of pure African ancestry, and about 85 for people of mixed European-African ancestry (mix ratio unknown, perhaps 50%). Comparing the 82 and 75 scores, one might infer that, over the past 300 years, the innate intelligence of African Americans has improved more than racial mixing can explain. It is not likely that more intelligent Africans were captured for deportation to the Americas; the opposite is more likely to have been the case. However, it is likely that the ocean voyage was an ordeal of which the more intelligent better coped, and therefore more likely survived. It is also likely that the more intelligent better coped with, and therefore more likely survived, the ordeal of adjusting to a life of bondage. Perhaps, the more settled life as bonded people on family farms facilitated the fatherhood of the more intelligent African American men. However explained, people of mixed African-European ancestry seem to have acquired a considerable boost in intelligence above that of their African ancestors. This result is also evident when examining societies in the Caribbean, and other nations outside of

Africa where people of African ancestry are in the majority. There, business and government leaders are predominantly people of a noticeable fraction of European ancestry, typically about one-half European ancestry.

Cognitive testing of East Asians, namely people from China (including Hong Kong and Singapore), Japan, and Korea, indicates that these people score slightly higher in intelligence than Europeans, probably about 106 versus 103. East Asians score especially high in spatial reasoning (perhaps 110), whereas they score much closer to Europeans in abstract reasoning (perhaps 98). This enhanced skill in spatial reasoning appears to explain the propensity for gifted Asian Americans to pursue careers in engineering, science, computing and medicine, instead of journalism and law. It is probable that Asians outside of the China-Japan-Korea region, (such as Mongolia, western China, Cambodia, etc.) would score lower than Europeans, perhaps more in line with Native Americans.

Intelligence testing of European Jews indicates that group scores exceptionally high, especially in verbal testing. These results indicate an average test score for European Jews of about 114. Herrnstein and Murray observed: "These test results are matched by analyses of occupational and scientific attainment by [European] Jews, which consistently show their disproportionate level of success, usually by orders of magnitude, in various inventories of scientific and artistic achievement."

Intelligence testing of Latinos, whose ancestry is at least 75% Native American, produces average scores of about 92, which is above today's African American average, and well below today's European American average. This indicates that the innate intelligence of the Native Americans who populated North America during early colonial days was about 87. That would have been about half way between the average for imported African slaves, at 82.1, and the Europeans, at 103.4.

Testing indicates a slight difference in the intelligence of males and females. European American males show a wider variation in test scores, and average about 2 points higher than females (105 versus 103). The wider variation in test scores means males are more prevalent than females at both the low end of the distribution and at the high end of the distribution.

There are strong correlations between cognitive ability and success in school and career. The greater an individual's cognitive ability, the more likely is his or her success in achieving high marks and advanced degrees in schools and universities. Similarly, the more likely is his or her success on a job that requires cognitive effort. Therefore, it is necessary to compare individuals of equal cognitive ability to judge today's opportunities in school and work for people of the three major American races. So, if we look only at people with cognitive ability that matches the European American standard of 100 – what I call "standard cognitive ability" – we find the following opportunities, by race.

1. African Americans of "standard cognitive ability" are more likely to hold a bachelor's degree (68%) than either European Americans (50%) or Latin Americans (49%).

2. African Americans of "standard cognitive ability" are more likely to hold an intelligence-oriented job, such as teacher, lawyer, physician, engineer, or scientist (26%), than either European Americans (10%) or Latin Americans (16%).

3. People of "standard cognitive ability" within all three races are being paid almost the same, the averages ranging from $25,001 for African Americans to $25,546 for European Americans (pay data for 1989).

4. African Americans are being given more opportunity to work as schoolteachers than are European Americans. The scores reported represent a mix of applicants and hired teachers – but all held teaching certificates from their respective college. Between 1983 and 1991, California teacher competency testing showed 80% of European American teachers passed, but only 35% of African American teachers passed. Testing in Georgia, between 1978 and 1986, showed 87% of European American teachers passed, but only 40% of African American teachers passed. A 1987 test in New York showed similar numbers (83% versus 36%). But, a 1989 Pennsylvania test was perhaps easier to pass (93% versus 68%). Although those teachers who failed these competency tests stand a better chance of not being hired or of losing their teaching job, many of those who failed remained on the payroll as working teachers. So, it is clear that the nation's school systems are giving African Americans more opportunity to work as classroom teachers than their cognitive qualifications would justify.

This concludes the presentation of my understanding of the distribution of cognitive ability resulting from the evolution of man from Creation to today's United States population. We now move on to the issue of Traits of Character.

Chapter 14. Racial Differences with Regard to Traits of Character.

There are strong correlations between cognitive ability and traits of character. People of every race with low cognitive ability are more likely to exhibit weak traits of character – such as lazy work ethic, criminal behavior, bearing children out of wedlock, and refusing marriage responsibilities. Therefore, it is necessary to compare individuals of equal cognitive ability to compare specific character traits of people within the three major American races. After selecting, for this comparison, a "standard cognitive ability" equal to the European American standard of 100, we find the following character traits by race using data from 1980 plus and minus a few years.

1. African Americans of "standard cognitive ability" are more likely to be living in poverty (11%) than European Americans (6%) or Latin Americans (9%); and, they are more likely to be unemployed for one or more months in a given year (15%) than European Americans or Latin Americans (both 11%).

2. African Americans of "standard cognitive ability" are less likely to be married (58%) than European Americans (79%) or Latin Americans (75%).

3. Female African Americans of "standard cognitive ability" are much more likely to bear a child out of wedlock (51%) than European Americans (10%) or Latin Americans (17%).

4. Female African Americans of "standard cognitive ability" are much more likely to have at some time been on welfare (30%) than European Americans (12%) or Latin Americans (15%).

5. African Americans of "standard cognitive ability" are much more likely to serve time in prison (5%) than European Americans (2%) or Latin Americans (3%).

These statistics were gathered within my lifetime, so they do not represent the environmental effects of long-ago African American bonding (slavery). Typically, the data is from about 1980, and that is 115 years, or 4-to-5 generations, after the last bonded African American was made independent (emancipated). It seems to me that these statistics show that African Americans are fortunate to be living in the United States, and to have access to so many opportunities. I invite you to compare these statistics to what you observe of people who live in Haiti.

A comparison between the opportunities available to people of African ancestry living in the United States, versus people of African ancestry living in Africa, is so dramatic that it is hard to describe in words. Today, people of African ancestry who live in the homeland are cursed with political genocide, hunger, and devastating disease – all horrors so grave that we Americans cannot comprehend them. And, life in Africa has become much more difficult since African people gave up European colonialism, and took control of the respective governments. History teaches us that, for Africans to live successfully, they need to be part of a society that is also populated with a large number of people of European or Asian ancestry.

Chapter 15. Racial Differences with Regard to Athletic Ability.

It is obvious to any objective observer of American athletic competitions that boys and girls, and men and women, with at least 25 percent African ancestry dominate sports teams where athletic ability is essential to successful competition. The best example is the game of basketball. In this game, athletic quickness and leaping ability are essential to outstanding play. Big, exceptionally tall European Americans can contribute to teams at center, and sharp-shooting European Americans with good ball-handling skills, and good play-making tactical leadership, can contribute to teams at guard. But, African Americans dominate at forward, and are overly represented at center and guard, as well. High school teams have a higher percentage of African American boys than junior high teams. The percentage increases in college teams. And, the percentage increases further in professional teams, where African Americans overwhelmingly dominate. Often, on professional teams, we see that all five players on the basketball court at any given time are of African ancestry. African Americans star on football teams, particularly where athletic skill is most important, most notably at wide receiver. African Americans dominate track and field events, especially at short-distance races. This is easily observed in American high schools and colleges. And anyone watching the Summer Olympics can readily see that people of African ancestry are dominant in the core track and field events. As an example, let me present a telling footnote:

The 200 meter men's sprint at the 2000 Olympics was won by Konstantinos Kenteris, a Greek man of European ancestry. He was the first man of non-African ancestry since 1972 to win an Olympic sprint in which the USA competed. At the conclusion of the race, Kenteris told reporters, "I knew I would be against 7 [runners of African ancestry] and wanted to do my best to wreck it for them. If it surprises you, that's good. It doesn't surprise me." A losing American runner, who was shocked as was most every sports observer, cried: "I can't believe he won." It took a cocky and blindly confident attitude, and some missed timing within the field of sprinters, for Kenteris to win, because men of African ancestry are clearly superior in that event.

But, who makes the most money each year, African American athletes at the top of their game, or European American business executives at the top of theirs? To find the answer we will simply look at employees, weeding out team owners and business owners.

Chapter 16. Comparison of Salaries: Superior Cognitive Ability, versus Superior Athletic Ability.

When we compare Americans with the greatest athletic ability to Americans with the greatest cognitive ability we find that, although the racial mix is totally different, similar salaries are paid. Let us look at individuals who earn $5,000,000 or more, annually. In the National Basketball Association 92 player-employees received 2001-2002 salaries equal to or greater than $5,000,000 each. Of these 92 men, 64 appear to be of three-fourths to full African ancestry, 15 appear to be from one-fourth to five-eighths African ancestry, one appears to be of perhaps one-eighth African ancestry, and 12 appear to have no African ancestry. The 64 players who appear to be of three-fourths to full African ancestry pulled down annual salaries that averaged $9,300,000 each. The 16 players who appeared to be of one-eighth to five-eighths African ancestry pulled down annual salaries that averaged $9,100,000, each. And the 12 players who appeared to have no African ancestry pulled down annual salaries that averaged $8,300,000, each. I compiled the above statistics from the NBA website during February 2002.

The Carolinas Panthers professional football team is dominated by African Americans. The quarterback is African American. A 2013 pictorial lineup of the starting defensive players showed 1 white man and 10 black men. A similar lineup of starting offensive players showed 3 white men and 8 black men. The 3-man squad of kickers was all white. The 3-man squad of returners was all black. I am going to forgo a presentation of player salaries, but rest assured, they are most generous.

At the same time, I also compiled statistics from the Forbes and Yahoo Financial websites, to determine the pay for top cognitive jobs, those being the chief executive officer jobs in American publicly held corporations. Here I eliminated CEO's who held sizable stock in the corporation for which they worked, because I wanted to look only at employees, not owners. I found 98 CEO's who were basically employees and who were paid total compensation, equal to or greater than, $5,000,000. The average total compensation of this group was $17,700,000, a figure that includes stock options. But, the average of their salaries and bonuses was only $5,300,000. None of these CEO's appears to have any African ancestry whatsoever. No person with any noticeable African ancestry had managed to win any of these 98 top CEO jobs·– not even one.

So, we find that athletic and cognitive salaries are similar at the top. But, because of the racial differences in athletic and cognitive abilities, African Americans dominate the top paying athletic jobs, while European Americans (whites), and a few Asian Americans, dominate the top paying cognitive jobs.

The above data shows how African Americans excel in the competitive struggle to rise to the top in the athletic arena, but not in the competitive struggle to rise to the top in the cognitive arena. Why is this so? The rewards of high pay for the most successful are similar in both fields, and, although African Americans have no "Affirmative Action" advantage in the athletic competitive struggle, they do have that advantage in the cognitive competitive struggle. The above outcomes were not caused by what today's African American political activists call "racism." The answer to this question goes back to the ancestors of Europeans and Asians, who were hunkered down in caves during long ago Ice Ages. By natural selection, in other words by Evolution, these cave-dwelling children lost a little athletic ability and acquired a little cognitive ability. The competitive struggles to reach the top athletic jobs, and the top cognitive jobs, simply, and naturally, result in basketball courts and football fields dominated by African Americans, and corporate chief executive offices dominated by European Americans and Asian Americans.

Chapter 17. Racial Differences with Regard to Musical Ability.

When Africans were first imported into the Americas, they arrived without demonstrable abilities to perform complex music. The musical instruments in use in sub-Sahara western Africa had been simple drums, simple flutes, and simple stringed instruments. The music they created was based on the more primitive 5-note scale. But, soon after their arrival, African Americans learned the European's more complex music scale, and began to excel. They sang in harmony. They learned to play European instruments. They composed. They eventually invented new musical forms, such as Jazz. As bonded people (slaves), and as independent people (emancipated), African Americans formed musical bands and played at dances and parties. At churches, African American men and women formed choirs, which were, in many cases, the best of their region. Obviously impressed, many European American musicians adopted African American musical forms and blackened their faces and hands with burnt cork to imitate them in staged minstrel shows. African American-inspired music, whether played by minstrels or by African Americans themselves, was very popular during the decade before The War Between the States. In fact, it was an Ohio-born member of a minstrel show who composed the compelling song, "I wish I was in Dixie's Land." That song, first performed in a minstrel show in New York City, became the most popular song among Confederate troops, who shortened the title to simply, "Dixie".

I first became aware of the tremendous musical talent of African Americans while a band member in high school. Nashville, Tennessee, schools were segregated back then, so my school, Hillsboro High, did not compete in athletics with the main African American high school, Pearl High. But we would meet at the annual Christmas parade downtown, and I immediately observed that the Pearl High band was far superior to the Hillsboro band. And, over the years, I have been uniformly impressed with African American church choirs whenever hearing them. Then recently, on a bareboat sailing trip in the eastern Caribbean, my sailing friends encountered the Union Island High School Steel Band. Here, dozens of students, most of pure African ancestry, displayed a great talent as they took turns at the steel pans. These students from one high school, on one poor and small island in the eastern Caribbean, under the leadership of one unimposing band director, were capable of thrilling visiting sailing enthusiasts from anywhere in America.

Today, African American musicians are more prominent in popular music than one would expect, considering their share of the American population. And the most popular African American musicians earn just as much money as the most popular European American musicians. On balance, it appears that the music ability of African Americans is at least equal to the music ability of European Americans, and in some ways might be superior. European Americans seem to be able to outshine African Americans only in the fields of country music and classical music.

Chapter 18. Summation and Thoughts about Your Conclusions.

My presentation is now complete. You now perceive my understanding. So let us together explore some pertinent questions and employ them as a way for you to arrive at attaining **your** understanding. As you review each question, reflect on your past understanding and beliefs and the basis upon which you have come to embrace each one. At the same time, please, allow scientific truths to trump so-called "politically correct truths." And listen to what we have before us handed down in our Bibles, keeping in mind that those words, those revelations by our Lord, were originally selected to match the level of scientific understanding of the first people to hear them, pass them to later generations and eventually commit them to writing. Again, we observe that God was not in the business of teaching science and geography to his people. Hopefully, the exercise below will broaden your mind, and enable you to acquire greater meaning from future endeavors. I am now through with the presentation of how the races of man came to be, and the story of God's six covenants with man, as, following Creation, he evolved over a span of about 200,000 years.

1. Did God create the Heavens and the Earth and its living creatures, using natural forces and evolutionary selection to advantage?

2. Do you agree that man is not just a highly evolved animal, that God, while employing supernatural powers, did select a site in East Africa about 200,000 years ago and there did create man in his own image? Did He – at the same time – give each person a soul capable of everlasting life?

3. Did God thereafter advance man through scientific principles of evolution and did He push a portion of mankind out of Africa about 110,000 years ago, and, by 30,000 years ago, had Ice Age hardships diversified man to include the European and Asian races?

4. By 15,000 years ago, had man eliminated all hominid animals that had preceded his arrival on Earth?

5. Was there, about 7,600 years ago, a catastrophic flooding of an advanced civilization around Black Lake, and did God establish a covenant with Noah at that time?

6. Did God later establish a covenant with Abraham, whose descendants became the Jewish people? Did He later establish a covenant with Moses, to prepare the Israeli people for nationhood?

7. Did God use the Jewish people as His witnesses, so, through them, He would become known to people from Egypt to Canaan to Persia to Rome, for the purpose of preparing the world for the arrival of His Son?

8. Then, during the days of the Roman Empire, did God make His New Covenant with all mankind through His Son, Jesus Christ? Did God, through Christ, command His disciples to spread the Gospel throughout the world?

9. Did Jesus Christ teach that "salvation is by faith, not by works, lest any man should boast"?

10. Should a Christian evangelize others by example, and by spreading the Gospel, but not by forcing his religious attitudes upon others? Is it possible for any religion to glorify itself by imposing its religious attitudes upon people who are not believers?

11. Does history show that, except for those venturing out of Africa about 110,000 years ago, Africans have otherwise been content to remain in Africa? In other words, was the African's historic reluctance to pick up and leave his homeland, more a timidity to challenge the unknown, than a passion to remain there? If Europeans and European Americans had not purchased bonded Africans (slaves), would there have been significant migration out of Africa during the last 500 years? Without forced migration out of Africa, thorough slavery, would there be a significant population of people of African ancestry in the Americas today? Did the importation of bonded Africans make possible their presence today in the United States? Are today's African Americans better off in every respect than their distant cousins who remain in the homeland and is that why essentially none wish to migrate to Africa? Is America fortunate to have many people of partial or pure African ancestry living among us throughout our nation, thereby making ours a nation of racial diversity not of racial purity? My answer to the final question is "yes". I hope yours is as well.

12. Did the European and Asian races, by being forced to overcome Ice Age cold by hunkering down in dark, but warm caves, lose athletic skills, but gain more helpful coping skills? Are much of the differences observed in the cognitive skills and athletic skills of individuals who are alive today rooted in their ancestor's different experiences in cold or temperate climates during past Ice Ages?

13. Did the Africans who came to the English colonies of North America contribute immensely to building our nation? Did the synergism of the African's compliant nature and athletic skills, when under the guidance of the European's coping skills, produce the most productive and powerful agricultural economy the world had known, up to The War Between the States and the mechanization of agriculture?

14. Do you see evidence that the method (slavery) by which Africans came out of Africa to settle in the Americas, although seemingly harsh and cruel, might have been God's way of working through man's nature to facilitate His objective of spreading the Gospel and building His harvest of souls?

Was slavery God's way of teaching European Americans to live alongside African Americans, and discover synergistic mutual benefits?

15. If essentially all Africans had, instead, been confined to their continent during the 1500's, 1600's, and 1700's, would that have worked greater hardship on scores of future generations? Did bringing Africans to America to live on and work on farms eventually open marvelous opportunities, and reveal the great contributions to mankind of which the African is capable?

16. Did God consider responsible owners of bonded African Americans (slaves) to have been His stewards over His flock of future believers? If so how does that relate to our relationships with people of other races?

I expect you know how I am answering these questions. What about you?

We are now arriving at the end of my presentation, *Understanding Creation and Evolution: A Biblical and Scientific Comparative Study*. If you found the last chapters with their emphasis on understanding racial differences an uncomfortable experience, I apologize. But, it is hard to truly present the subject of human evolution without digging into such details. Dr. Martin Luther King emphasized that people should be "judged by the strength of their character, not by the color of their skin." I totally agree; and it was toward measuring strengths of character that I constructed the final chapters of this study.

Keep in mind that you can gain knowledge from my study if you examine religious, or moral, or scientific, or political issues from a different perspective than I have. You do not have to accept all that I have embraced. Examine it bit by bit and embrace what you find helpful. Why?

Because, this study is irrevocably dedicated to my original objective: in all endeavors seek the truth – *"for the truth shall set you free."*

And I am hopeful you, too, will become a Truth-seeker.

The end.

References to this Study.

Note: The translation of the Holy Bible used throughout this study is the New International Version, copyrighted in 1978. I find it to be the most accurate translation and the easiest to understand among the translations available today and it is the one I have read from cover to cover.

The other references worthy of listing are given below:

1. *The Book of Man, The Human Genome Project and the Quest to Discover our Genetic Heritage*, Walter Bodmer and Robin McKie, Oxford University Press, 1994, 1997 paperback. This book and African Exodus below, when published, was a revolutionary window into how, for the first time in human history, examinations of the human genome was revealing, all around the world, human inheritance and racial ancestry.

2. *African Exodus, The Origins of Modern Humanity*, Christopher Stringer and Robin McKie, Henry Holt and Company, New York, 1996. The note above applies here as well.

3. *National Geographic Magazine*, July 2000, page 108. The sites at Dolni Vestonice and Pavlov are on the Dyje River, a tributary of the Danube, in present-day southern Czech Republic, about 700 miles upstream of what was Black Lake before the great flood. Here is amazing proof of early advanced civilizations upstream from Black Lake.

4. *The Southeastern Indians*, Charles Hudson, The U. of Tennessee Press, 1976, paperback reprint, 1994, pages 36 and 42. This explains how European diseased killed off almost all of the Natives in North America soon after Europeans arrived.

5. *Noah's Flood, The New Scientific Discoveries about the Event that Changed History*, William Ryan and Walter Pitman, Simon & Schuster, New York, 1998. Amazing. This discovery proving and dating the flooding of Black Lake, thereby creating the Black Sea, is fundamental to understanding the flood story in the Bible

6. *National Geographic Magazine*, April 1999. This gives details on long-ago human civilization at the time Black Lake was flooded by saltwater.

7. *The Charlotte Observer*, February 27, 2005. This reference is a remarkable and in-depth report on the Atlantic slave trade by reporters of the Charlotte Observer newspaper of Charlotte, NC. It was printed in their February 27, 2005 issue as a contribution to Black History Month. The story is presented as a full page color poster. The information was derived from several important studies of the Atlantic slave trade, with those sources properly noted.

8. *The Bell Curve, Intelligence and Class Structure in American Life*, Richard J. Herrnstein and Charles Murray, The Free Press, New York, 1994. This very important work by Herrnstein and Murray has been so bitterly denounced by so many vocal advocates of political correctness that you ought to suspect that it must be very important. That it is. It is both very important and also scientifically valid. I strongly support their work and rely on it. When you observe comments such as, "This study has been discredited," please ignore them and ask, "Show me a better study, likewise full of numbers and measures, that **improves** on the work of Herrnstein and Murray."

www.ingramcontent.com/pod-product-compliance
Lightning Source LLC
Chambersburg PA
CBHW080533030426
42337CB00023B/4707